An Analytical Approach to Optical Burst Switched Networks

T. Venkatesh · C. Siva Ram Murthy

An Analytical Approach to Optical Burst Switched Networks

Springer

T. Venkatesh
Department of Computer Science and
   Engineering
Indian Institute of Technology
Guwahati-781039
India
t.venkat@iitg.ernet.in

C. Siva Ram Murthy
Department of Computer Science and
   Engineering
Indian Institute of Technology
Chennai-600036
India
murthy@iitm.ac.in

ISBN 978-1-4899-8396-1   ISBN 978-1-4419-1510-8 (eBook)
DOI 10.1007/978-1-4419-1510-8
Springer New York Dordrecht Heidelberg London

Springer is part of Springer Science+Business Media (www.springer.com)

To The Almighty, my parents and my wife

Venkatesh

To the memory of my father-in-law, N.V.P. Sastry

Siva Ram Murthy

Preface

Optical burst switching (OBS) is envisioned to be one of the promising technologies to support all-optical switching in the core Internet. It combines the benefits of the fine-grained optical packet switching and the coarse-grained optical circuit switching while avoiding the limitations of both of them. In an OBS network, the data from the clients in the access network are collected at an ingress (edge) node, sorted based on the egress (destination) node, and grouped into variable-sized bursts. Increasing the basic unit of switching from a packet to a burst reduces the switching and processing overhead per packet. Prior to the transmission of the bursts, a control packet is sent to the destination node in order to reserve a lightpath for the corresponding data burst. The control packet is transmitted in the optical domain but processed electronically at each intermediate (core) node. The data burst is transmitted after some time without waiting for any acknowledgment from the destination node. By separating the control and the data planes in optical switching, OBS avoids the need for optical buffers as well as optical processing logic in the network. The packets are again extracted from the bursts at an egress node and then forwarded to their respective destinations in the access network.

OBS is finding potential application in grid computing and metro area networks to transfer bursty data across the Internet. In the recent past, significant research has been done to improve the basic architecture of the OBS and there have also been several testbed implementations across the world. One of the main problems with the bufferless switching in OBS networks is that the bursts are dropped due to the contention for the resources at the core nodes. Although several mechanisms have been proposed in the literature to reduce contention losses, they are not evaluated on a common platform making it difficult to compare them against each other. Research in the field of OBS networks requires strong analytical skills to understand the complex burst loss process. There are several factors that influence the blocking probability of bursts in these networks. The impact of each of these factors varies across different scenarios and cannot be understood only with simulations. Therefore, analytical models for the blocking probability help to understand the working principles of OBS in a better way. This book takes an analytical approach to study the traffic characteristics and blocking probability of the OBS networks.

This book serves as a good material to understand the analytical work done till date in the area of OBS networks. It classifies all the literature on modeling and analysis of these networks including the latest ones in four different chapters. Starting with the study of characteristics of bursty traffic, it discusses models for the blocking probability in different scenarios and the impact of contention resolution mechanisms on the blocking probability. Finally, this book studies the behavior of the transmission control protocol traffic over the OBS networks with appropriate models and discusses different proposals to improve its performance. Except for a few cross-references, this book is written in a way that a reader can understand each chapter independent of the others. It serves as a thought-provoking material for the researchers working on the analysis of high-speed networks. The scope of this book however is not limited to OBS networks alone but extends to high-speed communication networks with limited or no buffers. Since the research on high-speed networks with limited buffers is of great interest these days, this book is timely.

This book is mainly targeted at researchers working on analytical aspects of high-speed networks to provide them with a one-stop reference for different mathematical techniques used in modeling such networks. It is also useful for senior undergraduate or graduate students and the faculty teaching courses on the analytical aspects of high-speed communication networks. It also serves as a supplementary material to courses on modeling and analysis of communication networks.

By writing this book, we gained a number of ideas from numerous published papers and some books on this subject. We thank all those authors/publishers, too many to acknowledge individually. We would like to thank our institutions for providing us an opportunity to pursue this project. Venkatesh would like to acknowledge the financial support provided to him by Microsoft Research India with their Ph.D. fellowship. Siva Ram Murthy would like to acknowledge the research support he received from IBM Corporation through 2008 Real-Time Innovation Faculty Award. We would like to thank Raj Kumar who drew all the figures and the editor, Alex Greene, for his efforts in bringing out this book. Last but not the least, we acknowledge the love and affection from our families. This project would not have been successfully completed but for their understanding and patience.

Although reasonable care has been taken to eliminate any errors in the book, it would be great if the reader can find some time to send any suggestions to the authors.

Chennai, India   T. Venkatesh
August 2009   C. Siva Ram Murthy

Contents

Acronyms

AAP	adaptive assembly period
ACK	acknowledgment
AIMD	additive increase multiplicative decrease
ATM	asynchronous transfer mode
BACK	burst ACK
BAIMD	burst AIMD
BHP	burst header packet
BLE	burst length estimation
BLP	burst loss probability
BNAK	burst negative ACK
BS	burst segmentation
BTCP	burst transmission control protocol
BUPT	Beijing University of Posts and Telecommunications
CA	congestion avoidance
CBR	constant bit rate
CO	central office
CoCoNET	collision-free container-based optical network
CSO	constant scheduling offset
CTMC	continuous time Markov chain
ccdf	complementary cumulative distribution function
cdf	cumulative distribution function
DAP	dynamic assembly period
DBMAP	discrete-time batch Markov arrival process
DOBS	dual header OBS
DWDM	dense WDM
EFP	Erlang fixed point
ELN	explicit loss notification
FDL	fiber delay line

FGN   fractional Gaussian noise
FPGA   field programmable gate array
FR   fast retransmit/fast recovery
FTO   false time out

GAIMD   generalized AIMD
GGF   Global Grid Forum
GMPLS   generalized multi-protocol label switching
GRNI   grid resource network interface
GUNI   grid user network interface

HAM   heuristic aggregation method
HiTSOBS   Hierarchical TSOBS
HOBS   hybrid OBS
HSTCP   high-speed TCP
IBP   interrupted Bernoulli process
IP   Internet Protocol
IPP   interrupted Poisson process

iid   independent identically distributed

JET   just-enough-time
JGN   Japan Gigabit Network
JIT   just-in-time

LAUC   latest available unscheduled channel
LAUC-VF   latest available unscheduled channel-void filling
LOBS   labeled OBS
LSP   label switched path
LSR   label switched router

MAC   medium access control
MBMAP   min-burst-length–max-assembly-period
MONET   multi-wavelength optical networking
MSS   maximum segment size
mgf   moment generating function

NAK   negative ACK
NWF   nearest-wavelength-first

OBCS   optical burst chain switching
OBS   optical burst switching
OBT   optical burst transport
OCBS   optical composite burst switching
OCS   optical circuit switching

OFP	overflow fixed point
OLT	optical line terminal
ONU	optical network unit
OPS	optical packet switching
OXC	optical cross connect

PFB	per-flow-burst
PON	passive optical network
PPBS	probabilistic preemptive burst segmentation
PWA	priority-based wavelength assignment
P2P	peer to peer
pdf	probability distribution function
pgf	probability generating function
pmf	probability mass function

| QoS | quality of service |

RAP	resource allocation packet
RCDP	retransmission count-based drop policy
RPR	resilient packet ring
RTO	retransmission time-out
RTT	round trip time
RWA	routing and wavelength assignment
RdWA	random wavelength assignment

SACK	selective ACK
SAIMD	statistical AIMD
SCU	switch control unit
SIP	session initiation protocol
SOA	semiconductor optical amplifier
SOBS	slotted OBS
SONET	synchronous optical network
SRP	service request packet
SS	slow start
SynOBS	time-synchronized OBS

TAG	tell-and-go
TAW	tell-and-wait
TCP	transmission control protocol
TD	triple duplicate
TDM	time division multiplexing
TDP	triple duplicate period
TO	time out
TOP	time-out period
TSOBS	time-sliced OBS

UDP   user datagram protocol

WADM   wavelength add-drop multiplexer
WDM   wavelength division multiplexing
WROBS   wavelength routed OBS

Chapter 1
Introduction to Optical Burst Switching

1.1 Evolution of Optical Networks

The ever-increasing demand for higher bandwidth due to the growth of bandwidth-intensive applications such as remote information access, video-on-demand, video conferencing, online trading, and other multimedia applications motivated the search for alternatives to traditional electronic networks. Wavelength division multiplexing (WDM) is one such technology developed to handle the future bandwidth demands [52, 70]. In WDM networks, the huge bandwidth offered by a fiber is managed by splitting it into a number of wavelengths, each acting as an independent communication channel at a typical data rate of 10 Gbps. These networks are able to provide over 50 Tbps bandwidth on a single fiber. Characteristics such as low attenuation of signals (0.2 dB/km), extremely low bit error rates, i.e., fraction of bits that are received in error (10^{-12} in fiber whereas it is 10^{-6} in copper cables), low signal distortion, and free from being tapped (as light does not radiate from the fiber, it is nearly impossible to tap into it secretly without detection) are the additional advantages.

The first-generation optical networks which are currently deployed consist of point-to-point WDM links. Such networks are made up of several nodes interconnected by point-to-point links. At each node, data in the optical domain undergo conversion to the electronic domain for processing before being converted back into the optical domain. Such electronic processing leads to a considerable overhead if most of the traffic is bypass traffic and thereby the need for all-optical switching arose.

The second-generation optical networks consist of interconnected ring networks with each node equipped with wavelength add–drop multiplexers (WADMs) where the traffic can be added or dropped at each WADM node [70]. When the bypass traffic is high, WADMs help to reduce the overall cost by selectively dropping and adding traffic on some wavelengths and passing that on the other wavelengths untouched. In addition to the WADMs, these networks have optical cross-connects (OXCs) that are capable of switching traffic on any input port to any other output port. If required, these devices may be provided with wavelength conversion capability so that the traffic on one wavelength can be switched onto another wavelength

T. Venkatesh, C. Siva Ram Murthy, *An Analytical Approach to Optical Burst Switched Networks*, DOI 10.1007/978-1-4419-1510-8_1,
© Springer Science+Business Media, LLC 2010

on an outgoing port. With the development of such devices, switching data completely in optical domain is made possible.

The third-generation optical networks are based on a mesh network of multi-wavelength fibers interconnected by all-optical interconnection devices such as passive star couplers, passive routers, and OXCs. These networks are envisioned to have the capability of optical switching as well as reconfigurability. This generation of the optical network is envisioned to have two functional components: a network of optical switching nodes supporting all-optical data transmission and several access networks which feed the data to the core network through the edge (ingress and egress) nodes. The access networks could be based on the gigabit Ethernet, passive optical network (PON), or any other broadband access technology which is responsible to carry the traffic from the end users. The ingress nodes that connect the access networks to the core network are high-speed routers that can aggregate traffic destined to the same egress node. The core network consists of a mesh of reconfigurable elements (for example, OXCs) interconnected by high-capacity long-haul optical links.

There are three popular switching paradigms that have been proposed for use in the all-optical WDM networks, namely, optical circuit switching (OCS), optical packet switching (OPS), and optical burst switching (OBS) [70]. In OCS an end-to-end lightpath is established between a source node and a destination node for the entire session to avoid opto-electronic conversion at the intermediate nodes. In the futuristic OPS, packets from the access network are directly switched in the optical domain along the path which demands buffering and processing functions in the optical domain. The third switching paradigm OBS proposed to overcome the deficiencies of both circuit switching and packet switching is mainly targeted as a possible technology to support all-optical switching in the Internet before OPS is realized [85]. The benefit of OBS over conventional circuit switching is that there is no need to dedicate a wavelength for the entire session between a pair of nodes. Moreover, OBS is more viable than packet switching as the data burst does not need to be buffered or processed at the intermediate nodes. The OCS and OPS paradigms are briefly discussed in the following sections before a detailed introduction to OBS is provided.

1.2 Optical Circuit Switching

In OCS networks, dedicated WDM channels or lightpaths are established between a pair of source and destination nodes. A lightpath is formed by switching the traffic across the same or different wavelengths (if wavelength conversion is possible) on the links along a physical path. Such wavelength-routed networks consist of OXCs interconnected by point-to-point fiber links. The end user (end node) is connected to an OXC via a fiber link. The end node along with its OXC is collectively called as a node. Each node has transmitters and receivers (either tunable or fixed) for sending and receiving data on the lightpaths.

A lightpath must use the same wavelength on all the links in the route. This is known as the *wavelength continuity constraint* [70]. However, if wavelength converters are present, the lightpath may be converted from one wavelength to another at an intermediate node. A wavelength can also be used by multiple lightpaths as long as they do not have any common links along the route. This allows spatial reuse of wavelengths in different parts of the network and this property is called wavelength reuse. Wavelength reuse in wavelength-routed networks makes them scalable and cost efficient.

The establishment of lightpaths involves various phases such as topology and resource discovery, routing and wavelength assignment (RWA), signaling and resource reservation. Keeping track of network state information is the main goal of topology and resource discovery task. Network state information includes information about the physical network topology and the availability of the bandwidth on the network links. For wavelength-routed networks, information regarding the availability of wavelengths on a particular link in the network is very essential. One of the core issues in wavelength-routed networks is to determine the routes and assign wavelengths for lightpaths, known as RWA. Lightpaths may be set up in a static or dynamic manner. When the set of connections is known in advance, the problem of setting up lightpaths is known as static lightpath establishment and the goal is to minimize the network resources consumed such as number of wavelengths or fibers used in the network. In dynamic lightpath establishment, the lightpaths are set up in a dynamic way as the connection requests arrive in an online fashion. The objective in this case is to minimize the connection blocking and thus maximize the number of connections established in the network at any time. There is vast literature that deals with the problem of lightpath establishment for both static and dynamic cases [70].

One of the main disadvantages of OCS is that setting up a lightpath for long durations causes inefficient utilization of the resources particularly for bursty Internet traffic. A lightpath established once may remain so for days, weeks, or months during which there might not be sufficient traffic to utilize the bandwidth. To avoid such wastage of bandwidth, it is desirable that optical networks have the capability of switching Internet protocol (IP) packets directly and thus the paradigm of OPS evolved.

1.3   Optical Packet Switching

OPS is a switching paradigm that allows switching and routing of the IP packets in optical domain without conversion to electronic domain at each node. An OPS node has switching fabric which is capable of reconfiguration on a packet-by-packet basis. The optical packets are sent along with their headers without any prior setup into the network. At a core node, the packet is optically buffered using fiber delay line (FDL), while the header undergoes optical to electronic conversion and is processed electronically. Based on the header information, a switch is configured to transmit the optical packet from an input port to an output port after which it is released immediately.

Since network resources are not reserved in advance, some optical packets may contend for the same output port leading to packet losses. Lack of proper optical buffering technology aggravates the contention problem in OPS networks compared to traditional electronic packet switching networks, where electronic buffer technology is very mature. Optical buffering is realized through the use of FDLs which can hold an optical packet for a variable amount of time by implementing multiple delay lines in stages or in parallel [52]. The size of optical buffers is severely limited by physical space limitations. In order to delay an optical packet for a few microseconds, a kilometer of optical fiber is required. Because of this limitation, an OPS node may be very inefficient in handling high loads or bursty traffic. Another solution to the contention problem is to route the contending packets to a different output port other than the intended one. This technique is known as deflection routing. Deflection routing may cause looping and out-of-order delivery of packets and requires further study to tackle these limitations. Another issue in OPS is the lack of techniques for synchronization. In OPS networks with fixed length packets, synchronization of packets is required at the switch input port to reduce contention. Though some synchronization techniques have been proposed in the literature they are difficult to implement [9].

Practical deployment of OPS demands fast switching times while the optical switches based on micro-electro-mechanical systems offer switching times of the order of 1–10 ms. Though semiconductor optical amplifier (SOA)-based switches have considerably lower switching times (around 1 ns), they are quite expensive and the switch architecture uses optical couplers which results in higher power losses [59]. Apart from this there are also issues related to header extraction and optical processing that make the realization of OPS difficult in the near future [60]. To avoid the use of optical buffering and optical processing logic and still achieve switching in the optical domain, researchers came up with the OBS paradigm. OBS is considered a popular switching paradigm for the realization of all-optical networks due to the balance it offers between the coarse-grained OCS and fine-grained OPS [85]. Given that the data are switched all-optically at the burst level, OBS combines the transparency of OCS with the benefits of statistical multiplexing in OPS as described in greater detail below.

1.4 Optical Burst Switching

Figure 1.1a shows an OBS network with three components: an ingress node, an egress node, and a network of core nodes. At the ingress node, various types of client data from the access network are aggregated into a data burst which is transmitted entirely in the optical domain. Packets destined to the same egress node (destination edge node) and requiring the same level of service (if multiple classes are supported) are aggregated into a burst at a burst assembly queue. To avoid buffering and processing of the optical data burst at the intermediate nodes called core nodes, a control packet also called burst header packet (BHP), with the information about the length and arrival time of the data burst, is sent in advance. As shown in Fig. 1.1b, the BHPs are transmitted on a dedicated control wavelength, while the data bursts are sent after some time on separate wavelengths. The time lag between a BHP and the corresponding data burst is called the offset time which is set sufficient enough to enable the processing of BHP and to configure the switches at the core nodes. Once a burst reaches the egress node, it is disassembled into packets which are routed through the access network.

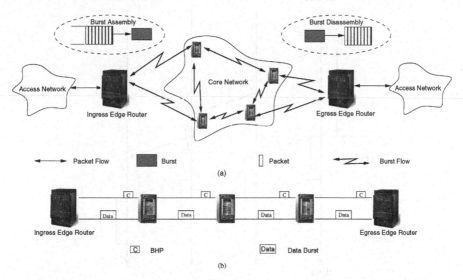

Fig. 1.1 (a) Architecture of an OBS network and (b) illustration of the transmission of BHP and data burst

Figure 1.2 shows a timing diagram to illustrate the offset time. Initially the offset time is set to T units of time which is calculated based on the number of hops to the destination node. Assuming that the time taken to process the BHP at a node is δ units, the offset time is set to $h.\delta + t_s$, where h is the number of hops and t_s is the time required to configure the switch. During the δ units of time, a BHP is processed in the electronic domain to find out the requirements of the corresponding data burst and to find a feasible schedule for it. The switches along the path are configured only when the data burst arrives (in t_s units of time) to enable the data burst to cut through an all-optical path. At the egress node, the data burst is disassembled into IP packets. Generally it is assumed that the data bursts are buffered only at the edge nodes in electronic domain since the technology for optical buffers is immature. This protocol for wavelength reservation in which the wavelength is reserved only for the duration of the data burst and the reservation of the resources along the path is not acknowledged in any way is known as just-enough-time (JET) protocol [85]. When a data burst arrives at the core node and finds all the wavelengths busy at that time it is simply dropped. Such losses are termed as *contention losses* which mainly occur due to the simultaneous arrival of bursts exceeding the number of available wavelengths at that instant. Handling contention losses is a problem unique to bufferless OBS networks, and it cannot be handled in the same way as congestion in traditional networks (which occurs due to overflow of a queue) for the simple reason that these losses are temporary in nature and may not be due to the lack of bandwidth.

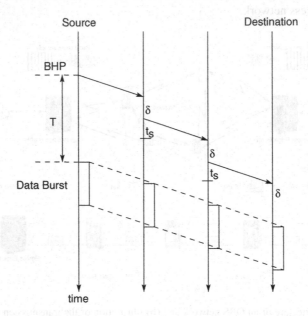

Fig. 1.2 Illustration of offset time

OBS has the advantages of both OCS and OPS while overcoming their shortcomings. A brief comparison of the three switching paradigms based on various factors is given below [85]:

- *Bandwidth utilization*: Link utilization is lowest in OCS networks. A dedicated lightpath is set up for every pair of the nodes which cannot be used by the passing traffic even when the load is low. OCS networks cannot switch traffic at a lower granularity than a wavelength. But OPS and OBS networks allow traffic between multiple node pairs to share the bandwidth on a link due to statistical multiplexing.
- *Setup latency*: OBS networks use one-way signaling schemes for reserving resources on the path before the data are transmitted. Setup latency is very low in such networks, unlike OCS networks where dedicated signaling messages are exchanged between the source and destination nodes to set up and release the lightpaths.
- *Switching speed*: OPS networks require very high speed switches to switch optical packets which are small in size. On the other hand, switching is faster in

OCS networks. Further, lightpaths are usually set up for longer durations, and therefore the time for configuration of the switches can be longer. In the case of OBS networks, medium-speed switches are sufficient due to the larger size of the optical bursts compared to the optical packets.

- *Processing complexity*: In OPS networks, as the control information is contained in the optical packet, the processing complexity is very high. The header has to be extracted from each packet and processed electronically. In OCS networks, as the lightpaths are of a longer duration, the complexity is relatively low when compared to OPS and OBS networks. Because of the larger granularity of bursts which are made up of several IP packets, the processing complexity of OBS falls in between that of OCS and OPS.
- *Traffic adaptivity*: OCS networks are not adaptive to variable bursty traffic due to the high setup latency and the use of coarse-grained wavelength switching. But OPS and OBS networks are capable of handling bursty traffic as they support statistical multiplexing.

In what follows, a brief introduction to the different components of an OBS network, major research issues in this area, and finally, some applications and testbed implementations of OBS are presented.

1.4.1 Node Architectures

Figure 1.3 gives a block diagram that shows the functions of different nodes in an OBS network. There are primarily two types of nodes in an OBS network: the edge nodes (ingress and egress nodes) and the core nodes. An ingress node collects the IP packets from the access network into bursts and also generates the control packets which are responsible for reserving the resources along the path. The main functions of an ingress node are burst assembly, signaling, generating the BHPs, determination

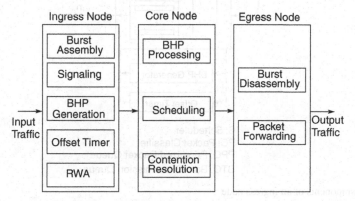

Fig. 1.3 Functional diagram of nodes in OBS network

of the offset time, and RWA. The core nodes are responsible to schedule the data bursts arriving on an input port to an output port based on the information provided by the BHPs. The core nodes are also responsible for resolving the contention among multiple bursts. The egress node disassembles the bursts into IP packets and forwards them to the access network.

Figure 1.4 shows the main components of an OBS ingress node. The IP packets from the access network arriving at the ingress node are classified based on the destination address by the packet classifier. The packets are sent into the burst assembler module where a separate queue is maintained for each destination node and for each class of service. Each burst assembly queue generates bursts with all the packets destined to the same egress node. There may be separate packet queues

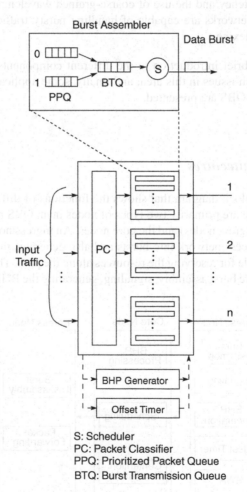

S: Scheduler
PC: Packet Classifier
PPQ: Prioritized Packet Queue
BTQ: Burst Transmission Queue

Fig. 1.4 Components of an ingress node

in the burst assembler for each class of service (two queues are shown in the figure corresponding to classes 0 and 1). Once the bursts are created they are scheduled for transmission on the outgoing link by the scheduler. The ingress nodes have electronic buffer used for the burst assembly and scheduling. Different burst assembly techniques are used to aggregate the packets into bursts based on a threshold for either time, burst size, or both. Once the burst is assembled, a BHP is generated by the BHP generator module which carries the information required for scheduling the burst along the path. There is another module to determine the offset time (shown as offset timer in the figure) based on the number of hops to the destination node.

Figure 1.5 shows the architecture of a core node. It consists of a switch control unit (SCU) and an OXC. The SCU configures the OXC and maintains the forwarding tables. When the SCU receives a BHP, it consults the routing and signaling processors to identify the output port based on the destination. If the output port is available at the time of data burst arrival, the SCU configures the OXC to let the burst pass through. In case the port is unavailable, SCU configures the OXC based on the contention resolution policy implemented. In case the data burst arrives at the OXC before the BHP it is simply dropped. This problem is known as *early burst arrival* which might happen due to an insufficiency in the offset time. The SCU has many important responsibilities which include BHP processing, burst scheduling, contention resolution, forwarding table lookup, switching matrix control, header rewrite, and controlling the wavelength converters. At the physical layer of the OBS, there is a lot of scope for research in the area of design of high-speed burst-mode transmitters, receivers, and optical switches.

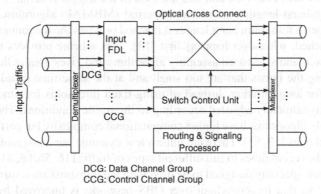

DCG: Data Channel Group
CCG: Control Channel Group

Fig. 1.5 OBS core node architecture

1.4.2 Burst Assembly

Burst assembly is the process of assembling data bursts from IP packets at an ingress node of the OBS network. An important factor in the burst assembly is the trigger criterion to determine the time when a burst is released. The trigger criterion changes the characteristics of the traffic in the network. It also determines the arrival process of the traffic and thus the loss process. The burst assembly algorithm should not delay the packets for a long time to create larger bursts and at the same time should not lead to the formation of several smaller bursts. The burst assembly algorithms mostly use a threshold on either the time or the burst size or both to trigger the release of a burst. In time-based assembly, a burst is sent into the network at periodic time intervals which are of variable sizes depending on the arrival rate. It gives uniform gaps between successive bursts from the same ingress node. In time-based assembly, the length of the bursts depends on the arrival rate at the ingress node [22]. In size-based assembly, fixed-size bursts are generated at non-periodic time intervals.

Each of these methods of burst assembly suffers from some shortcomings at low or high traffic load because it considers only a single criterion for the burst formation. Under the low load condition, the size-based assembly does not provide an upper bound on the time to wait before releasing the burst. In such conditions, if the size threshold is large, the average waiting time of the packets in the assembly queue becomes large. Similarly, under heavy load conditions, the time-based assembly leads to large burst sizes and the average waiting time becomes larger compared to that in the case of size-based assembly. So under heavy load conditions, size-based assembly is preferable while time-based assembly is preferable under low load conditions. Clearly, it is desirable to have a burst assembly algorithm that performs well under all traffic load conditions which leads to the design of mixed assembly algorithms in which both time and size are used as the trigger criteria.

In the min-burst-length–max-assembly-period (MBMAP) algorithm, a burst is generated when a minimum burst length is reached or when the maximum assembly period is reached, whichever happens first [13]. This scheme provides the best of both time-based and size-based assembly algorithms and gives greater flexibility. It avoids sending the bursts that are too small and at the same time avoids delaying the packets for long duration. Instead of using fixed thresholds for time and size, they can be dynamically adjusted according to the traffic conditions. The dynamic burst assembly algorithms have higher computational complexity but perform better than the fixed ones [13, 81]. There are quite a few dynamic burst assembly schemes proposed in the recent times to suit different types of traffic [16, 56, 66, 81]. Some of them have been specially designed to be used with the transmission control protocol (TCP) traffic so that its throughput over OBS networks is improved by reducing the burst size in case of congestion. Some of them use prediction techniques to determine the arrival rate of the traffic at the edge node and transmit the BHPs in advance thereby avoiding the delay due to the burst assembly [39]. Analyzing the characteristics of the assembled traffic shows that the assembly algorithms play a major role in shaping the traffic in the OBS network. Some analytical models for the traffic in OBS network are discussed in Chapter 2 which show that the traffic becomes smoother after the burst assembly while the basic characteristics of the input traffic remain intact. Traffic smoothing due to the burst assembly was observed to simplify the traffic engineering and capacity planning in OBS networks [22].

1.4.3 Signaling Schemes

There are two types of protocols mainly used for the reservation of resources along the path namely tell-and-wait (TAW) and tell-and-go (TAG) [85]. In TAW, the BHP first travels and reserves all the resources along the path to the destination. If the reservation is successful, an acknowledgment (ACK) message is sent back by the destination, after which the data burst is sent. If the reservation fails at any intermediate node in the path, a negative ACK is sent back to the source to release the resources at the upstream nodes. In TAG protocol, the source does not wait for an ACK from the destination. The BHP first travels and reserves bandwidth on the links, but the data burst is sent after the offset time even before the BHP reaches the destination. If the BHP fails to reserve the bandwidth on any link, the data burst is simply dropped or an appropriate contention resolution policy is considered when it reaches that link. JET and just-in-time (JIT) are the two main signaling schemes based on the TAG protocol. In JIT, the wavelength on which the data burst is due to arrive is reserved immediately after the BHP is processed and is torn down using explicit control messages. Although the BHP and the data burst travel on different wavelengths, the BHP needs to inform the core node only about wavelength on which the data burst is scheduled to arrive. In JET, the bandwidth is reserved only for the duration of the data burst. When the BHP arrives on the control wavelength, it carries information about the burst length, arrival time, and the wavelength used. The wavelength is scheduled for use only after the offset time for the duration of the burst. This improves the utilization of the wavelength but increases the processing time of BHP at the node. The timing diagram of burst transmission with JET protocol is shown in Fig. 1.2. In JET protocol, there is a possibility of the data burst being blocked by another burst that is scheduled to arrive later because the offset time varies according to the path length. This phenomenon is known as *retro-blocking* due to which the bursts with larger offset time are successful in reserving the wavelength before the bursts with smaller offset time. Although the variable offset time is useful in providing service differentiation between the bursts, it increases the blocking probability of bursts with smaller offset time, i.e., those that travel shorter paths [18]. Since the wavelength is reserved even when there is no burst, the utilization in JIT is poorer compared to JET and the blocking probability is also higher. JET protocol outperforms JIT protocol in terms of bandwidth utilization and even the burst loss probability at the expense of increased computational complexity at the core nodes. However, JIT is useful to serve the traffic that requires guaranteed service.

1.4.4 Routing and Wavelength Assignment

One of the important problems with the OBS networks is the loss of bursts due to contention among simultaneously arriving bursts at a core node. Mechanisms to reduce contention can be implemented either at the core nodes or at the edge nodes.

Contention resolution at the core nodes requires additional hardware or software components thereby increasing the cost of deployment significantly. At the edge nodes, contention losses can be effectively countered by path selection (routing) and wavelength assignment [93]. These techniques minimize control signaling, are scalable, and are cost-effective for deployment. Traffic engineering with appropriate path selection and wavelength selection strategies has received considerable attention but requires additional work to reduce the time taken for route discovery.

Routing on a per-burst basis causes delay in burst transmission in addition to the assembly and processing time. However, if all the ingress nodes select shortest paths to the egress nodes, a bottleneck might be created at some core node leading to contention losses. Switching between alternate paths to an egress node was shown to reduce the contention losses to a large extent [17]. In OBS networks, this strategy is referred to as path selection. A set of pre-computed link-disjoint paths are available at each ingress node and the path used for transmission is changed periodically based on the observed burst loss probability (BLP). A suite of path switching algorithms, classified as pure and hybrid schemes, were proposed in [93]. Route discovery in OBS networks based on any cost metric considering the wavelength continuity constraint or other constraints can be done using the generalized multi-protocol label switching (GMPLS) suite of protocols. Once a pre-computed set of paths is given, the best path at any given time can be selected based on a performance measure and subsequently the optimal wavelength might be chosen. Careful selection of wavelength at the edge node was shown to yield a low BLP even without the use of any other contention resolution mechanisms [77]. Several adaptive and non-adaptive wavelength assignment heuristics to select a wavelength that minimizes the BLP were proposed. Some of the proposed heuristics were seen to improve the BLP by as much as two orders of magnitude compared with the simple first-fit and random wavelength assignment strategies [77]. There are also some proposals to handle the path selection and wavelength assignment in a joint fashion using machine learning techniques [79, 80].

1.4.5 Channel Scheduling

When a BHP is processed at the core node, a feasible schedule is identified for the data burst. Since the scheduling is done in an online fashion based on the arrival time of the burst, efficient scheduling algorithms are necessary to minimize the burst losses. Depending on the availability of wavelength converters, the possible wavelengths for scheduling the burst on the outgoing link are identified. The main objective in the online scheduling in OBS networks is to minimize the voids, where a void on a wavelength is the idle time between the transmission of two consecutive bursts. Several scheduling algorithms have been proposed for OBS networks in which the main objective is to reduce the BLP along with improving the utilization by minimizing the voids [37, 86]. One of the scheduling algorithms which is very simple and used widely is the latest available unscheduled channel (LAUC) scheduling algorithm also known as *Horizon* [92]. The algorithm maintains the latest available unscheduled time on each wavelength and when a new burst arrives it is scheduled on a wavelength such that the void size is minimized. For example, in Fig. 1.6 the wavelengths W_2 and W_3 are the unscheduled ones at time t but W_2 is selected to transmit the data burst because $t - t_2 < t - t_3$. Simplicity and ease of implementation are the main advantages of the LAUC scheduling algorithm as the scheduler needs to remember only one value for each data wavelength which is the unscheduled time. But the LAUC scheduling algorithm cannot utilize the voids created by the previously scheduled bursts because it looks at only the latest unscheduled time. For example, it does not consider the void on W_1 because it is considered to be already scheduled up to $t_1 > t$. This causes excessive burst losses when the variation in the offset time is large.

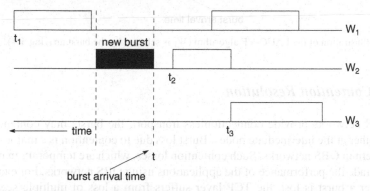

Fig. 1.6 Illustration of the LAUC algorithm (W_2 is selected by the burst arriving at t)

The void filling version of the LAUC scheduling algorithm, the latest available unscheduled channel-void filling (LAUC-VF) scheduling algorithm, keeps track of all the voids on the wavelengths and tries to schedule a burst in one of the voids whenever possible. In Fig. 1.7, at the time of the burst arrival at t along with W_5 being considered an eligible unscheduled wavelength, the remaining ones are also considered because the voids are sufficiently large to schedule the data burst at t. The LAUC-VF scheduling algorithm selects the wavelength with the minimum value of $t_i - t$, for $i = 1, 2, \ldots, 5$. W_1 is chosen to carry the data burst as $t - t_1 < t - t_i$, for $i = 2, 3, 4, 5$. The LAUC-VF scheduling algorithm is more complex to implement than the LAUC scheduling algorithm because it has to maintain additional state information for each wavelength which is the size of the available voids. More details on the LAUC-VF scheduling algorithm can be found in [92]. Since the LAUC-VF scheduling algorithm can use the voids created by previously scheduled bursts, the link utilization is higher than that in case of the LAUC scheduling algorithm. However, the LAUC-VF scheduling algorithm takes longer time to schedule a burst compared to the LAUC scheduling algorithm. A few other burst scheduling algorithms developed recently have lower complexity while some of them are designed to consider physical impairments on the wavelengths [14, 21].

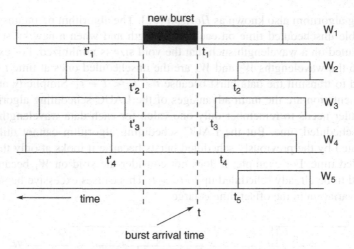

Fig. 1.7 Illustration of the LAUC-VF algorithm (W_1 is selected by the burst arriving at t)

1.4.6 Contention Resolution

As OBS networks provide connectionless transport, the bursts may contend with one another at the intermediate nodes. Burst loss due to contention is a major source of concern in OBS networks. Such contention losses, which are temporary in nature, can degrade the performance of the applications using OBS networks. For example, whenever a burst is lost, the TCP layer suffers from a loss of multiple segments which could all be from the same congestion window. In such a case, depending on the size of the congestion window TCP sender might even go to the time-out state. Repeated burst losses lead to very low throughput and result in wastage of resources due to retransmissions at the TCP layer. These effects are studied in greater detail in Chapter 5.

Contention among two bursts occurs due to their overlap in time when they arrive simultaneously on two different links or wavelengths and request the same out-going wavelength. In electronic packet switching networks, contention is handled by buffering. However in optical networks, buffers are difficult to implement and there is no optical equivalent of random access memory. So when multiple bursts contend for the same wavelength at a core node, all but one of them are dropped. The contention loss is illustrated with an example network with 10 nodes shown in Fig. 1.8. The first figure shows two bursts originating at nodes 1 and 7, respectively. Assume that both the bursts are destined toward the same destination node, say node 6. Contention occurs when the bursts that arrive on the links 4–5 and 10–5 request the same wavelength on the link 5–6. Assuming that there is no wavelength converter at node 5, only one of these two bursts can be scheduled at a time and the other one is dropped. The decision to drop the burst is actually made before the data bursts arrive at node 5 after processing their BHPs. In such a case, the data burst that arrives later is dropped at the core node. There is no way of communicating this information to the ingress node and the data burst cannot be delayed more than the offset time at the ingress node. There is no feasible way to send a negative ACK to signal the loss of a burst due to contention. Since a burst consists of several packets, its loss causes serious degradation in the performance of the higher layer protocols. Hence, design of contention resolution schemes to reduce the BLP is the focus of research in OBS networks.

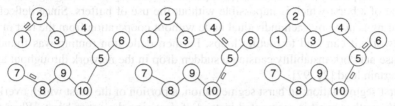

Fig. 1.8 Contention loss in OBS networks

Some of the main techniques used to resolve contention for a wavelength at the core nodes are mentioned below and also discussed in detail in Chapters 3 and 4.

- Optical buffering: Optical buffers based on FDLs can be used to delay the bursts for a fixed amount of time [15]. Optical buffers are either constructed in single stage, which have only one block of delay lines, or in multiple stages which have several blocks of delay lines cascaded together, where each block consists of a set of parallel delay lines. Optical buffers can be broadly classified into feed-forward, feedback, and hybrid architectures [52]. If a delay line connects the output port of a switching element at one stage to the input port of another switching element at the next stage it is called feed-forward architecture. In a feedback architecture, the delay line connects the output port of a switching element at one stage to the input port of a switching element at the previous or the current stage. A hybrid architecture combines both feed-forward and feedback architectures. More recently, using both FDLs and wavelength converters, a feed-forward-based architecture is used to implement either a shared or a multiple-input buffer on the input side such that it remains scalable with increasing number of ports [94].
- Wavelength conversion: If two bursts contend for the same wavelength at a core node, they can be sent on different wavelengths to resolve the contention. Wavelength conversion is the process of converting the bursts on one wavelength in an incoming link to a different wavelength in the outgoing link. This helps to increase the wavelength reuse and the wavelengths can be spatially reused to carry different bursts on different fiber links in the network. Wavelength conversion is of four types: full wavelength conversion, limited conversion, fixed conversion, and partial wavelength conversion. In full conversion, any incoming wavelength can be shifted to any outgoing wavelength, while in limited conversion, not all incoming wavelengths can be translated to all the outgoing wavelengths [61]. In fixed conversion, each incoming wavelength may be translated to one or more predetermined wavelengths only. In partial wavelength conversion,

different nodes in the network can have different types of wavelength conversion capability [3].

- Deflection routing: This is a technique of deflecting the bursts onto alternate paths toward the destination in case of contention for a wavelength at a core node [88]. Deflection routing has many disadvantages. It is by nature suboptimal, since it only considers the congestion of the current switch, not the state of links in the deflected path. Deflection routing implementation requires changing the offset time of a burst which is impossible without the use of buffers. Since deflection routing is done by each individual node without cooperation from the rest of the network, it can lead to routing loops. Further, deflection routing was found to cause network instability causing a sudden drop in the network throughput after a certain load [27, 97].

- Burst segmentation: In burst segmentation, a portion of the burst which overlaps with another burst is segmented instead of dropping the entire burst. When two bursts contend for the same wavelength, either the head of the contending burst or the tail of the other burst is segmented and dropped. Therefore segmentation can be classified into head dropping or tail dropping. The remaining segment of the burst is transmitted successfully to the destination thereby increasing the packet delivery ratio. A combination of both segmentation and deflection routing has also been shown to be effective in reducing the BLP [85].

1.4.7 Service Differentiation

Although it is desirable to adapt the quality of service (QoS) solutions proposed in the Internet to OBS networks, it is difficult due to lack of buffers at the core nodes. Without buffers at the core nodes, the loss performance of different classes of traffic heavily depends on the characteristics of the traffic from the edge nodes. Longer bursts with shorter offset times have a higher probability of being dropped than the others. Hence, even within one class it is difficult to achieve consistent performance if the burst length and offset times are variable. There are several approaches to provide service differentiation in OBS networks which can be applied either at the edge nodes or at the core nodes [17]. The architecture of the ingress node shown in Fig. 1.4 supports QoS provisioning for the aggregated traffic according to the requirements at the higher layers. Since there is a separate burst assembly queue for each class of service node and destination node, the burst assembly algorithm can be different for each class of service.

A simple method of providing QoS support at the edge node is to use variable offset time for each class of traffic. The bursts with larger offset time always succeed in reservation and have a lower probability of getting blocked. The blocking probability of different classes when offset-based service differentiation mechanism is used at the edge node is discussed in detail in Chapter 3. It is observed that by using sufficient variation in the offset times between two classes of bursts, the higher priority class can be isolated from being blocked by the lower priority class.

The performance of the extra offset time scheme for service differentiation can be improved by the use of FDLs though it is not mandatory. However, this scheme increases the end-to-end delay apart from being unfair to low-priority traffic. It also reduces the utilization of the bandwidth. Wavelength preemption techniques are used at the core nodes to give prioritized treatment to different classes of bursts in which the lower priority bursts are preempted by the higher priority ones when there is contention. Prioritized segmentation can also be used to provide service differentiation. Class-based aggregation with prioritized burst segmentation is yet another technique used to provide QoS guarantees [75, 84].

Another technique to provide service differentiation in JIT-based OBS networks is the burst cluster transmission [74]. In this technique, a mixed assembly algorithm is used to form a cluster of bursts which are scheduled with a cluster scheduling algorithm such that the low-priority bursts are placed ahead of the high-priority ones. The bursts are picked up from different queues and arranged in increasing order of the priority into a cluster. The burst cluster is scheduled as a whole and routed through the path set up by a single BHP. At any core node, the low-priority bursts from the cluster are dropped till the resources are free. In this way, the control overhead is reduced albeit with a slight increase in the delay at the ingress node and the low-priority bursts are differentiated from the higher priority ones based on the loss rate. In addition to dropping the low-priority bursts till the contention is resolved, the probabilistic preemption approach can be used to improve the delivery rate of higher priority bursts. A combination of burst cluster transmission and probabilistic preemption technique was shown to be effective in supporting differentiated service in OBS networks [73]. Various techniques proposed to support relative and absolute QoS in OBS networks have been classified based on whether they are end-to-end techniques or node-based techniques and discussed in detail already in [17]. Some of them have been discussed in Chapters 3 and 4 to understand their effect on the blocking probability.

1.4.8 TCP over OBS

Despite the proposal of several contention resolution mechanisms for OBS networks, providing reliable transport of bursts without the use of explicit reservation of wavelengths is a challenging task. Further, contention losses in OBS networks are fairly common depending on the traffic pattern and the topology so that they do not necessarily indicate congestion in the network (indicated by unavailability of bandwidth). A burst loss can cause several TCP/IP packets to be lost which affects the performance of many Internet applications that use the TCP for transport. Analyzing the behavior of the TCP traffic over OBS networks is challenging due to many reasons. The two main factors that affect the performance of the TCP over the OBS networks are the burst assembly process and the inherent bursty losses. Due to this the behavior of a TCP flow over an OBS network varies significantly compared to that in the Internet. First, assembling TCP packets at the ingress node, disassembling them at the egress, and the buffering at the core nodes (if used) increases the round-trip time (RTT). Next, the burst losses which may lead to the simultaneous loss of multiple segments from a window lead to fluctuations in the evolution of the congestion window of TCP. Such losses also cause TCP to identify congestion falsely even when it does not exist. The effect of these factors on the throughput of a TCP flow is elaborated in Chapter 5. Several mechanisms are proposed to improve the performance of TCP over OBS networks which either use contention resolution techniques at the core nodes to reduce the BLP or use explicit feedback on the nature of the loss to avoid the reduction of congestion window for every burst loss. Some of the variants of TCP proposed for OBS networks modify the congestion control mechanism to differentiate between the two types of losses in OBS networks, i.e., congestion and contention. The other proposals either adapt the burst assembly mechanism or adjust the rate of the traffic at the edge node to reduce the BLP and thus improve the performance of TCP. For a detailed survey on the techniques used to improve the performance of TCP over OBS networks and different variants of TCP over OBS networks, the reader may refer to [68].

1.4.9 Other Issues

Apart from the major issues discussed above, there are a few other issues that have received considerable attention in OBS networks which are briefly discussed here. In OBS networks, similar to the OCS networks, multicasting is achieved through light splitting which inherently results in power loss. It is important to note that the dynamic nature of reservation in OBS networks makes it suitable for optical multicasting because the resources of the multicast tree are reserved on a per-burst basis. However, setting up multicast sessions in OBS networks poses significant challenges due to the burst contentions apart from the power losses at the splitter nodes. At the core nodes, an incoming burst is split into multiple bursts at the optical level using splitters which are then sent to the multicast group of nodes. The number and location of these splitters should be considered while designing efficient multicast protocols for OBS networks. A suite of multicasting protocols have been proposed for OBS networks which are investigated in terms of the control overhead and the BLP. There are two main approaches followed to set up multicast sessions in OBS networks. In the first approach, traffic from a multicast session is assembled into multicast bursts at the ingress node. For each multicast session a separate queue is maintained and the bursts are sent over the multicast tree constructed with any of the multicast protocols. Each ingress node establishes independent trees for the multicast sessions. A multicast session can also be treated as multiple unicast sessions and the bursts from multicast traffic are assembled along with the unicast bursts so that the control overhead is minimized. In the second approach, multicast trees set up by an ingress node are shared by traffic from other nodes also. In this approach the average burst length would be longer than the case without tree sharing. The control overhead is further reduced in this case since only one BHP is sent for multiple sessions and the resources along the multicast tree are released immediately after each burst is transmitted [30, 31, 78].

Protection of paths from the failure of links or nodes along the path is another issue that is important in WDM networks. Similar to OCS networks, $1 + 1$ path protection scheme for OBS networks does not require any additional functionality at the core nodes. In this scheme two disjoint paths are computed for each ingress–egress node pair. If the signal quality degrades or a link is broken on one of the paths, the traffic is switched to the other path automatically. Apart from the regular protection switching schemes, segment protection scheme that uses deflection routing with burst segmentation was also found to be very effective to provide protection against failures [23]. As soon as a failure is detected, the upstream core node deflects the bursts destined to the affected output port on an alternate path. If burst segmentation is also used as a contention resolution mechanism, deflection routing is used in conjunction with it to reduce the packet losses due to network failures.

1.5 Variants of OBS

Several variations to the basic JET/JIT-based OBS architecture have been proposed to reduce the burst losses due to contention and improve the performance of the network. This section presents a few of the major variants of OBS, some of which have also been discussed in greater detail in [17, 49]. Although there are many such variants which completely eliminate the contention losses, none of them has attracted attention like the basic JET-based architecture due to its low complexity in the implementation. Hence, the focus of this book is only on the JET/JIT-based OBS networks.

1.5.1 Labeled OBS

The JET-based OBS network architecture mainly decouples the processing in the control plane from that in the data plane and avoids the need for optical buffers or processing logic. However, these networks use a dynamic one-way signaling mechanism that leads to contention losses. In such a network, routing either unicast or multicast sessions along with the contention resolution mechanisms such as deflection routing makes traffic engineering a challenging task. If there is any additional delay in the processing of the BHP at a core node, the data burst might be dropped due to insufficient offset time. Therefore, designing efficient signaling protocols for the control plane becomes a challenging task to handle. The GMPLS framework has many protocols for routing, signaling, and traffic engineering which are used effectively to solve the problems in OBS networks. This variant of OBS enabled with a GMPLS-based control plane is known as labeled OBS (LOBS) [57].

In LOBS, each OBS node is augmented with a GMPLS-based controller much similar to the label switched router (LSR) in the conventional GMPLS networks.

Each BHP is sent as an IP packet with a label in the control plane over a pre-established label switched path (LSP). The corresponding data burst can also be sent along the same LSP used by the BHP. By using the explicit and constrained-based routing capabilities of the GMPLS framework, routes can be set up according to the requirements of the traffic which simplifies the traffic engineering in OBS networks. An established LSP can use the same or different wavelengths along the path depending on the wavelength conversion capability. An additional advantage in OBS networks compared to the conventional GMPLS networks where multiple LSPs cannot be merged onto a single path is that LOBS allows bursts belonging to multiple LSPs to use the same path. This is because the BHPs that are processed electronically contain the details about the data burst unlike the label in GMPLS networks which only determines the route. At the edge LOBS router, label stacking and LSP aggregation can also be performed along with the aggregation of IP packets so that multiple LSPs can be multiplexed over a single path. The LOBS architecture also supports the protection of paths by either the $1 + 1$ scheme or any other shared-path protection scheme by establishing redundant LSPs during the BHP routing phase. Such schemes require no additional actions from the core nodes in the OBS network at the time of recovering from a failure and also speeds up the restoration [23, 57].

1.5.2 Wavelength-Routed OBS

Wavelength-routed OBS (WROBS) combines the advantages of the two-way reservation mechanism in circuit-switched networks with the JET-based OBS architecture to avoid the contention losses. WROBS uses a centralized scheduler to allocate wavelengths to requests originating at the ingress nodes. When a burst is assembled at the ingress node, a request is sent to the centralized scheduler which searches for a common (multiple) free wavelength(s) on all the paths to the egress node. If a lightpath is found, the scheduler sends an ACK to the ingress node with the wavelength information so that the burst can be sent over the wavelength specified without any contention. Since a centralized scheduler processes all the requests for reservation, a centralized RWA can be used to identify an optimal lightpath. The end-to-end delay and loss can be guaranteed in this framework. The reservation of resources is done only for the duration of the burst which improves the network utilization compared to the OCS networks [19]. WROBS also allows the requests to specify end-to-end delay and packet loss rate. Since it assumes a fast circuit-switched end-to-end lightpath establishment with guaranteed and deterministic delay, it supports dynamic changes in bandwidth requests. To provide QoS in this architecture, the centralized controller can use different RWA algorithms and prioritized scheduling mechanisms for each class of traffic [34].

   However, this solution is not scalable due to the dependence on the centralized scheduler for each burst transmitted. The control overhead is very high since there are two control packets sent to an ingress node and one control packet sent to each core node for each burst. Further, there is an idle time before the wavelength is identified and the burst is transmitted which increases with the network size reducing the network utilization. To improve the utilization when the network size is large, wavelength reservation is proposed to be done in a proactive manner using prediction techniques for the distribution of traffic requests [33]. Instead of the request for wavelength reservation being generated at the end of the burst formation, it is sent as soon as the first packet arrives at the burst assembly queue. The control packet for the scheduler in this case has projected information about the burst length, waiting time at the assembler, and the arrival time of the burst. Using these predicted values, the scheduler assigns an appropriate wavelength for a duration slightly larger than the burst length to account for any additional delay. This method parallelizes the burst assembly and scheduling so that the network utilization is better than that in WROBS.

The centralized and distributed control architectures for wavelength reservation in OBS networks, namely, WROBS and JET/JIT-based OBS, have their own advantages and disadvantages. In the distributed reservation model, design of efficient techniques to compute the offset time, scheduling the resources, and fault management are challenging tasks due to the lack of global knowledge about the network resources. On the other hand, while centralized control simplifies these issues it leads to issues related to scalability as the network size increases. In order to resolve the deficiencies associated with both the architectures, a hybrid architecture for signaling and scheduling is proposed in [60]. In the hybrid architecture, the entire network is divided into domains or clusters using a clustering technique and each cluster has a scheduler to handle the requests pertaining to that domain. In the hybrid architecture each domain operates in a centralized manner while the inter-domain communication is handled in a distributed fashion. Each domain also has a secondary scheduler that handles the requests if the main scheduler fails. To support QoS for different classes of traffic, both one-way and two-way reservation mechanisms are supported with the low-priority bursts transmitted using the one-way reservation mechanism. A JET-based signaling protocol is used across the domains and the offset time is computed according to the estimated time for communication across the domains.

1.5.3 Dual-Header OBS

This variant of OBS is designed to reduce the delay incurred by the BHP at each node by decoupling the resource reservation and the scheduling operations. In OBS each BHP incurs a variable processing delay at each node which reduces the offset time as the burst travels through the path. Due to the variation in the offset time, there is unfairness to the bursts traveling shorter paths. To maintain a fixed offset time for all the bursts traveling through the paths with different lengths, dual-header OBS (DOBS) uses two different control packets for each burst [7]. The first one is the service request packet (SRP) which contains the information necessary to schedule the bursts, namely, offset time, length of the burst, and class information. The second one is the resource allocation packet (RAP) which is used to configure the switch. Unlike the conventional OBS, the burst is not scheduled as soon as the control packet arrives. First, the SRP is processed to determine the resources required for a burst and then it is forwarded to the next node without waiting for the burst to be scheduled. The burst is scheduled after a certain time just before the data burst arrives at the node. As soon as the burst is scheduled, the RAP is sent to the downstream node to indicate the wavelength selected for the burst. After the RAP is received from an upstream node, the burst is scheduled using the information in the RAP and the SRP received earlier.

Similar to the conventional OBS, the time gap between the burst and the SRP is reduced as the burst travels along the path but the offset time which in this case is the time between the arrival of the data burst and the actual time at which it is scheduled can be adjusted using the RAP. Since the order in which the incoming bursts are scheduled at a node is a function of the offset time, DOBS completely controls the order in which the bursts are scheduled. Each node can select its own scheduling offset (time between the arrival of the burst and the time at which it is scheduled) to account for the differences in the switching speed. DOBS allows the core nodes to precisely control the offset times without the use of any FDLs and thereby allows the core nodes to control the QoS parameters. The main benefit of the DOBS architecture is the constant scheduling offset (CSO) where each link may be assigned an independent CSO value to ensure that all the bursts can be scheduled in a first-come-first-serve basis. CSO-based DOBS has a lower scheduling complexity compared to the JET/JIT-based scheduling and has better network utilization compared to the JIT signaling. DOBS achieves better throughput-delay performance compared to either JIT or JET signaling because the resource reservation and scheduling are parallelized. Further, DOBS has better fairness compared to the JET signaling because voids are not created during the scheduling which reduces unfairness to bursts of larger length. It also avoids path length-based unfairness because it maintains a constant offset time for all the bursts traveling along the path [8].

1.5.4 Time-Slotted OBS

The slotted-time variants of OBS switch bursts in the time domain instead of the wavelength domain to avoid the need for wavelength converters at the core nodes [58]. In time-sliced OBS (TSOBS) a wavelength is divided into periodic frames each of which is further subdivided into a number of time slots. The data burst is divided into a number of segments with each segment having a duration equal to that of the time slot. The length of the burst is measured in terms of the number of slots it occupies. Each burst is transmitted in consecutive frames with each segment of the burst using the same slot in every frame. Each incoming link is assumed to have a synchronizer to align the boundaries of the slots with the switch fabric. The BHP needs only to indicate the arrival time of the first segment of the burst, the position of the time slot in the frame, and the number of slots required to transmit the burst. If all the frames have free slots in the required position, then the burst is transmitted; otherwise, it is delayed using the FDLs for the required number of slots. The maximum delay that can be provided by the FDL is kept same as the maximum number of time slots in a frame. A burst is dropped if it cannot be scheduled within the maximum delay possible. To improve the BLP in TSOBS, the burst scheduler can scan all the available time slots for each incoming burst and choose the set of slots which minimizes the gaps between the newly scheduled bursts and the existing ones. This is similar to the LAUC-VF algorithm for the conventional OBS in which the probability that the future bursts can find higher number of time slots is maximized [29].

Another slotted-time variant of OBS similar to TSOBS is the time-synchronized OBS (SynOBS) which is proposed to minimize the use of FDLs as well as the wavelength converters. In this architecture, several FDL reservation mechanisms are studied for their blocking probabilities. With the discrete-time Markov chain-based analysis, the optimal size of the time slot which reduces the blocking probability was found [62]. The BLP in the case of TSOBS and SynOBS for a multi-fiber network with different classes of traffic was analyzed to find that the performance of these variants is same as the conventional OBS with wavelength conversion [38].

To reduce the BLP in TSOBS, explicit wavelength reservation can be used before a burst is transmitted but it increases the signaling overhead. Another proposal, optical burst chain switching (OBCS), combines the merits of both explicit reservation and one-way reservation [44]. In OBCS, multiple non-contiguous and non-periodic bursts on a wavelength termed as *burst chain* is switched together using explicit reservation so that the control overhead is minimized while retaining the advantages of TSOBS. Based on the measured arrival rate of bursts between an ingress and egress node pair, the bandwidth demand is predicted at the beginning of each frame. The availability of time slots on every wavelength is collected using the probe packets. Once the probe packet collects the information about the free time slots on each wavelength, the ingress node compares it against the predicted bandwidth information. Based on the probed information, the ingress node keeps updating the prediction algorithm which is used to reserve the time slots for future bursts. An explicit reservation request is sent along the path to reserve the time slots before a burst chain is sent to avoid any losses.

Recently, another slotted-time architecture called the slotted OBS (SOBS) is proposed which completely eliminates the need for optical buffers and wavelength converters [104]. SOBS uses a synchronizer at the edge node which eliminates the randomness in the burst arrival and thereby losses due to contention. It also creates bursts of equal length so that the wastage of bandwidth due to the voids is minimized. Since all the bursts are of same length and their departures are synchronized they can be scheduled at the beginning of the frame. If a burst needs to be dropped because all the slots are busy, then it can be dropped completely. Theoretical analysis of SOBS shows that the utilization of the link and the burst delivery ratio are much better than that in the conventional OBS much similar to the differences between

ALOHA and slotted ALOHA. Along with these basic design parameters, SOBS also controls the rate at which the bursts are released from the ingress node and supports load balancing algorithms to minimize the possibility of burst overlap. Following the proposal of SOBS and TSOBS, there has been considerable amount of work to improve the efficiency of slot allocation with centralized controller and to improve the fairness and the utilization [43]. In general the time-slotted variants of OBS have been shown to provide better QoS guarantees for the traffic compared to the conventional OBS.

While TSOBS and the other related variants of OBS are successful in reducing the burst loss even without wavelength conversion, they have rigid frame structure. The size of the slot and the frame, i.e., the number of slots in a frame, has been observed to be the main factors to determine the BLP in TSOBS. A small frame size increases the probability of contention while a large frame size causes large end-to-end delays because each flow has access only to a small portion of the available bandwidth. This loss–delay trade-off is fixed for any class of traffic because the frame size is fixed across all the flows. To overcome this limitation and to provide differentiated service in terms of the loss–delay characteristics, the frame structure is designed to be flexible in hierarchical TSOBS (HiTSOBS) [71]. HiTSOBS allows frames of different sizes to co-exist together in a hierarchy such that the slots higher in hierarchy offer service to higher data rate traffic. The delay-sensitive traffic is supported by frames higher in hierarchy where the frames are of smaller size. At the same time, the frames lower in hierarchy can support loss-sensitive traffic. Along with the ability to support different classes of traffic simultaneously, HiTSOBS also allows dynamic changes in the hierarchy of the frames according to the mixture of traffic classes thus obviating the need for any other changes in the network. Similar to the TSOBS, a BHP carries the information about the number of slots required to transmit the burst as well as the level at which the burst has to be transmitted. However, in HiTSOBS the bursts are scheduled atomically rather than slice-by-slice to serve the entire burst in a frame at the desired level. This also keeps the control plane scheduling scalable and reduces the number of operations in the data plane to the number of levels in the frame hierarchy.

1.5.5 Hybrid Architectures

OBS networks are suitable mainly for short-lived flows or bursty traffic because they provide better bandwidth utilization with statistical multiplexing of the traffic. But they are not appropriate for continuous stream of data because they increase the delay due to assembly/disassembly at the edge nodes and require frequent configuration of the switches. Hence, to provide good service to both bursty and continuous streams of traffic, several hybrid architectures have been proposed which combine burst switching with circuit switching and flow switching. Each edge node in these architectures contains two modules: one to handle the bursty flows and another to handle continuous stream of traffic. The incoming traffic is either classified into best-effort traffic and guaranteed traffic or short-lived and long-lived traffic. Since the traffic in the Internet today has data, voice, and video traffic (real time or stored), this classification is appropriate. The short-lived flows (best-effort traffic) are handled with OBS protocols while the guaranteed traffic is handled with OCS or any other two-way reservation protocol [36, 90]. In addition to combining OBS and OCS, virtual circuit switching is also supported by the hybrid architecture to support traffic with stringent QoS requirements [5].

To avoid contentions in an OBS network a few of the architectures proposed the use of a centralized scheduler similar to WROBS. One such architecture proposed to avoid the contentions in the optical domain and to eliminate the optical processing in the core network is collision-free container-based optical network (CoCoNet) [51]. The main objective in the design of this network is to eliminate optical processing in the core network and thus reduce the transmission delay which is one of the major components of end-to-end delay. The network is designed as a ring topology physically but forms a mesh at the logical level to enable centralized scheduling of requests. The channel is slotted and the transmissions are synchronized across different links. Another level of aggregation is introduced above the optical layer where the data are aggregated into containers. The containers allow aggregation of different types of traffic from the access networks. The containers are not inspected at any core node but simply forwarded to the destination. At the ingress node, a synchronous slotted transmission scheduler is used for signaling and data aggregation. The aggregation of data into containers ensures guarantees to all kinds of traffic either short-lived or long-lived. The containers are designed in such a way that they can be routed effectively across multiple domains without being examined at the boundaries. The optimal container size is determined based on the data rate, utilization of the links, and the network. It was shown through analysis and implementation that it is indeed possible to enable zero-loss bufferless switching in the core networks with appropriate aggregation, synchronization, and scheduling strategies [51].

Another hybrid architecture called hybrid OBS (HOBS) uses the idle time during the establishment of the lightpath in two-way reservation to send the low-priority bursts [83]. HOBS combines one-way with two-way reservation under a single unified control plane for QoS differentiation. In WROBS, the reservation of resources is done with a two-way signaling which has an idle time significantly high due to large propagation and scheduling delay. A control packet is used to set up the resources along the path which are reserved only when an explicit ACK is sent back along the path. In HOBS, the same setup message is used to identify the resources for both the low-priority and the high-priority bursts. While the high-priority bursts are sent using the WROBS signaling mechanism, the low-priority bursts are transmitted by the time the reservation is acknowledged. Using the same control packet for both the low- and high-priority bursts reduces the signaling overhead. The setup message carries the necessary information about both the categories of bursts. The HOBS signaling acts like a TAG protocol for the low-priority bursts and TAW for the high-priority bursts. The switch architecture is also modified appropriately to handle both signaling protocols. It was shown that HOBS can achieve good data transmission rate for both classes of bursts with a bounded end-to-end delay.

1.6 Applications of OBS

OBS is endowed with the capabilities to satisfy the high bandwidth demands of applications such as grid computing, IP television, and cloud computing. OBS is still in the nascent stage of development but is expected to take off rapidly once it is commercialized by a leading industry participant. In spite of a lot of research activity in the area of OBS networks, the technology is yet to find its way into deployment beyond controlled laboratory environments. One of the reasons for this was found to be lack of efforts to standardize various OBS protocols and architectures. Focusing on developing products for specific applications can also speed up the deployment of this technology [49].

1.6.1 Grid Computing

In the recent times, a working group in the global grid forum (GGF) is committed toward standardization of OBS in the context of grid computing [69]. Existing architectures based on optical networks for large-scale grid computing applications mainly proposed the use of dedicated lightpaths to transfer the job and its associated data to the sites where the required resources are present [54]. While the use of dedicated lightpaths for grid computing is acceptable for scientific applications that demand the transfer of large data sets, it is not so for certain applications which fall in the class of consumer grids. Applications such as multimedia editing, interactive gaming, and online visualization of virtual environments demand a consumer-oriented grid which is basically a grid service accessed by several users dynamically which has different requirements compared to the scientific grid computing applications [46]. As long as the data sets in the application remain large, it is advantageous to use OCS where one or more complete wavelengths are used for the data transfer. However, if only small- or medium-sized data sets are to be transferred and processed as in consumer grids, it is desirable to use bandwidth at the sub-wavelength level. OBS becomes a natural choice for consumer grids due to many reasons. Before the reasons for the popularity of OBS in consumer grid applications are mentioned, the characteristics of consumer grids need to be understood.

A typical consumer-oriented grid application might be a multimedia editing application in which integrated audio and video manipulation programs allow the users to manipulate video clips, add effects, or render the films. Advances in the technology for recording, visualization, and effects will demand more computational and storage capacity which may not be locally available at each user. In consumer grids, even the residential users connected over a low bandwidth broadband access network are given direct access to the computing power of the grid. Every user working on bandwidth-intensive applications cannot be allowed to establish a lightpath for each operation. Since the transmission times are of the order of microseconds, wavelengths cannot be set up for each operation which requires switching time in the order of milliseconds. A large user base and variety of applications give rise to highly dynamic service requests for the grid which means that a dedicated static infrastructure is not the optimal solution. Further, the holding time of the requests is much smaller than the setup time. The requests are also large in number so that scheduling them must be handled in a decentralized fashion for faster response time.

Due to the above-mentioned characteristics of consumer grids, OCS is not an optimal technology for use in such applications. OBS allows decentralized operation of the network with a capability to handle large number of requests in a dynamic fashion. Assuming that each job along with the associated data can be transmitted in a single burst, the wavelength is reserved only for the duration of the transfer and the path is established with a control packet. Since the blocking probability of requests becomes negligible with large number of wavelengths, OBS can be used in the consumer grids enabling bufferless optical transfer. Since no explicit setup of the paths is needed, the response time experienced by the user is reduced drastically compared to that in the OCS network. OBS allows the use of optical transport without wastage of bandwidth yet allowing a large number of users to access the grid. Hence OBS is widely accepted as a suitable technology for bandwidth-intensive consumer grid applications [12, 48].

Figure 1.9 shows a typical architecture of the network to support consumer grid applications with OBS. The job requests are generated by the client applications on the fly and each request is collected along with its necessary data set into a single optical burst (typically in the size of megabytes). The burst is preceded by a BHP which contains the characteristics of the job that includes the required resources. The grid user network interface (GUNI) is designed to transport the pair of BHP and the burst along the optical network while the grid resource network interface (GRNI) is designed to enable the access of computational and storage facilities in the network. This architecture has the potential to offer global access to the computational and storage resources in the network for a large number of users with different profiles. One of the main advantages of this architecture is that both traditional data traffic and grid traffic can be supported simultaneously with a common infrastructure. All OBS routers are capable of forwarding the bursts with conventional protocols and also support intelligent transport of grid traffic. An OBS router in this architecture should be capable of providing QoS support to the traffic as well as have the capability to route anycast requests [20]. Since the grid requests do not have the destination address, several anycast routing protocols have been proposed for OBS networks in the context of grid computing [45, 47]. The main function of the GRNI is to handle the signaling between resource managers and the optical network. In addition to this it also handles advanced reservation, propagation of resource state, service-related events, flexible bandwidth allocation and returns the results to the source or alternate destinations. Similarly, GUNI performs functions such as flexible bandwidth allocation, claim existing agreements, automatic scheduling and provisioning of requests, and traffic mapping into bursts.

In the light of grid computing applications, there are new challenges to be handled in the design of OBS networks. Handling reliability is an important issue which ensures the integrity of the data and minimizes the wastage of resources. Since the

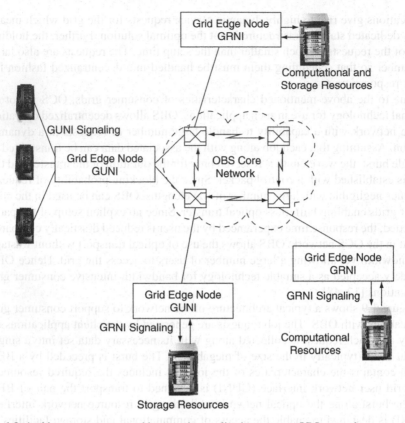

Fig. 1.9 Architecture to support grid computing applications over an OBS network

destination site is not fixed a priori protocols such as the TCP cannot be used to provide reliable transfer of the data [87]. New signaling protocols are required to transport the control messages between the entities in the network which include storage and computation sites apart from the source node. Minimizing the control messages is an important issue of research given the dynamic nature of the requests [4]. Design of new routing protocols is another issue of importance. Along with the main function of forwarding the bursts between the core routers, routing protocols should also ensure timely delivery of the job to the corresponding resource and the result back to the source or an alternate destination. Good anycast routing protocols can make the end users capable of resolving the requests to suitable destination in a flexible way [24]. Discovering the service and reserving the resources is another important issue in this context [98].

Instead of using the OBS network only to transport the job and the data sets, designing networks that are application-centric is another popular strategy in the recent times. Creating an application-aware OBS network needs both an efficient transport network and an application layer signaling protocol that can help users indicate their needs. An application-aware OBS network also helps the resource providers to advertise the services and resources to the users. A popular approach to design user-centric OBS network for grid computing is the design of session initiation protocol (SIP)-based OBS network [10, 12]. In this approach, the SIP is used as a layer between the grid-specific control messages and the network control plane. The proposed architecture was implemented in an existing OBS network testbed and was successfully tested. Three different approaches, namely, overlay, partially integrated and fully integrated, were also proposed to implement SIP-based OBS network and the merits and demerits of each of them are investigated in detail [11, 101]. The integration of SIP with the OBS control plane enables the optical network to use application layer information to establish and manage connectivity between the users and the resources or applications. Such a network was found to be effective in managing and negotiating application sessions in conjunction with the physical layer connections.

1.6.2 Access and Metro Networks

Although OBS is proposed as a technology for the all-optical core network, there have been few proposals to use OBS in the access and metro networks as well. The main idea behind this is that currently the physical transport in optical access networks uses circuit switching while the traffic is predominantly bursty. The resilient packet ring (RPR) technology standardized as IEEE 802.17 is another technology that aims to improve the bandwidth efficiency by allowing spatial reuse and differentiated bandwidth provisioning based on class of frames. To allow flexible resource management of the optical fiber in the access network, OBS is also proposed to be used as an alternative medium access control (MAC) layer as well as an access network technology [28, 32, 67].

Since the traffic in the Internet is predominantly bursty in nature, optical burst transport (OBT) is proposed as a technology even for the metro area networks. OBT uses burst-mode transmission between the senders and the receivers on a WDM ring topology and avoids the complexity of electronic processing. The burst-mode transmission is made possible by swift reconfiguration of short-term lightpaths across the optical switches interconnected in a ring topology. One wavelength is used for control signaling while the others are used for the data transmission. A token-based access protocol is used in OBT where the control signals include the token, the BHP, and other network management messages. Due to the separation of the control and data planes, control signal for a destination can be transmitted while the data for another destination is being transmitted without any collision. The size of the data burst is kept variable to allow flexibility between the two extremes of data-oriented and circuit-oriented services for the optimal use of resources. In order to manage multiple data channels, multiple tokens can be simultaneously propagated over the same control channel to carry access grants for respective data channels. Combined with tunable transceivers, multiple tokens can manage the WDM network without an increase in the number of transceivers. Since OBT can access the data channel at a time through a dedicated token, the network does not have collisions on a single data channel. To improve the resource utilization OBT allows spatial reuse of wavelengths. Compared to the other access technologies such as synchronous optical network (SONET) which use signaling mechanisms like GMPLS to realize dynamic bandwidth provisioning with configuration times of the order of seconds, OBT enables sub-millisecond times for provisioning. A prototype of OBT is also developed to demonstrate the performance of the proposal, the results of which showed that it is compatible with the Ethernet technology and can ensure faster response time for video transport [32].

To bridge the gap between the high-speed optical transport network at the core and a relatively slower metro network, several proposals have come up in the recent times. Prominent ones among them are the RPR, long-reach PONs, and hybrid PONs which all lack an all-optical access–metro interface which results in a bottleneck called the metro gap [49]. OBS is used to improve the utilization of the bandwidth in the access network as well as to create an all-optical access–metro interface [67]. In the access networks such as the PONs, OBS is proposed to be used to transmit information in the form of bursts over the lightpaths connecting the central office (CO) to the optical network unit (ONU) at the end users and vice versa. This enables the sharing of optical line terminal (OLT) at the CO among all the ONUs. In the OBS-based PON architecture, a centralized scheduler is used at the OLT for the MAC protocol which combines WROBS and TDM for the downstream and upstream data burst transmission in the dynamic bandwidth allocation protocol. Since the propagation time is much lower than the burst aggregation time in the access networks, WROBS is used in the OBS PON architecture. Further, WROBS also avoids collisions in the network. Apart from using the OBS in the access network, a new OBS multiplexer node is designed with arrayed waveguide grating multiplexers and wavelength converters to interface with the metro network. To reach long distance routers, the OBS multiplexer node acts as a metro node as well as an OLT for the ONUs. Unlike the earlier proposals which used OBS either in the access network alone or in the core network with an optical burst add/drop interface to the metro network, this architecture enables the extension of the access network to a metropolitan reach [67].

1.7 Testbed Implementations

To date there have been several attempts to test the various architectures and protocols in OBS networks in a controlled laboratory environment around the world. Although these experiments have been carried out only with small-sized networks, they serve to demonstrate the practicality of the proposals. A few of these implementations evaluate the protocols in basic OBS architecture like deflection routing and RWA while the others evaluate some of the variants of OBS like the LOBS. However, building a real OBS network environment to evaluate any kind of architecture or protocol is still an open issue. An ideal testbed is expected to emulate real OBS network as far as possible, provide a flexible network-wide performance evaluation platform for new ideas and novel technologies, and contribute a viable solution to the next-generation optical Internet [96].

A software prototype of the JIT protocol was successfully demonstrated in the year 2000 under the multi-wavelength optical networking (MONET) program in which an asynchronous transfer mode (ATM)-based control was used to set up and tear down the connections [89]. Earliest demonstration of a prototype of the OBS router dates back to the year 2002 when a high-speed switch matrix based on SOA was integrated with ATM interfaces used to emulate burst assembly and disassembly functions [50]. An implementation of the JIT protocol on an OBS network testbed with a core node and three edge nodes where the control packets are retransmitted when they are lost demonstrated the advantages of using this protocol for signaling [91]. JIT protocol was also extensively used as a part of the JumpStart project which led to the implementation of several OBS networking testbeds along the east coast of the USA. JumpStart supported ultra-fast provisioning of lightpaths by implementing JIT signaling in the hardware at the intermediate nodes and allows the end users to set up and tear down the lightpaths [6]. The testbed developed in the JumpStart project was used to realize dynamic establishment of lightpaths for sending uncompressed high-definition television signals, grid applications, file transfers without the need for transport layer and low-latency, zero-jitter supercomputer applications.

Another testbed developed at the University of Tokyo is used to implement a mixed assembly algorithm that classifies packets based on the destination node and QoS parameters to support multiple classes of service [72]. Along with the basic functionalities, the advantage of using deflection routing and priority-based wavelength assignment (PWA) was also demonstrated. With PWA, an edge node increases the priority of a given wavelength if the bursts have been sent successfully on this wavelength. Otherwise, the priority of the wavelength is decreased. The edge node always assigns a wavelength with the highest priority value in order to reduce the BLP. PWA is discussed in detail in Chapter 4. Experimental results showed that both PWA and deflection routing are effective in reducing the BLP but deflection routing was found to be more effective than PWA. A few major projects funded at the University of Tokyo led to the development of several advanced optical switching devices such as switch matrix, tunable lasers, and wavelength converters. Using these devices, a prototype for the OBS node was also built to demonstrate the advancement of technology in this area [53]. On similar lines, a high-speed switching matrix capable of attaining switching times in the order of microseconds was developed and used to construct an OBS node capable of bit-rate-independent wavelength conversion. This node was then used to build a three-node network in which a dynamic deflection routing scheme was also implemented [1]. Experimental results comparing the losses with and without deflection routing demonstrated that near-theoretical low frame loss rate is achievable in layer-2 transport over OBS networks.

The Japan gigabit network (JGN) II is another testbed developed over the years and is widely used to test several OBS protocols. OBS using a two-way signaling protocol based on the GMPLS was demonstrated to show that it is suitable for metro area networks with acceptable round-trip propagation delays [3, 63]. A congestion-controlled OBS network that provides connection guarantees to the bursts as short as 100 ms with switching time within 10 ms in metro area networks less than 200 km long was designed and implemented on JGN II [65]. Given a bound for the BLP, the network is designed in such a way that at any point of time the BLP is not more than the desired value. Using PWA and deflection routing, the BLP was always maintained to be within the required value. The JGN II testbed was also used to demonstrate 40 Gbps transmission of optical bursts with an error rate less than 10^{-6} over a distance of 92 km between two hops [76]. An optical label recognition scheme was implemented in field programmable gate array (FPGA) board which is then integrated with an OBS node prototype that can potentially be scaled up to 64×64 ports. An error-free optical burst forwarding over more than 10-hop network under variable length, asynchronous burst arrival process was demonstrated. A variable bit-rate optical burst forwarding with contention resolution by wavelength conversion was also implemented [2]. With these experiments the coordination of OBS with upper-layer protocols was also demonstrated by implementing an Ethernet-frame-to-optical-burst converter edge node, which successfully transmitted real application data over the OBS network at 40 Gbps.

Several experiments have also been conducted in the testbed developed at the Beijing University of Posts and Telecommunications (BUPT) in China. A four-node star network was built to implement the JET protocol along with the LAUC-VF scheduling algorithm from which the significance of void filling algorithms in improving the BLP compared to the non-void filling ones was studied [26]. Supporting applications with different QoS requirements by varying the offset time were also demonstrated in the same testbed [25]. A testbed with two edge nodes and three core nodes was developed to demonstrate the possibility of implementing time division multiplexing (TDM) in OBS network. To this extent, the JET protocol was extended to support TDM services and support both periodic and aperiodic traffic [95]. A key feature of the extended JET is the integration of signaling with the reservation mechanism. The one-way reservation mechanism of JET was extended to two-way reservation to provide QoS guarantees to the periodic traffic. The extended JET was implemented in the testbed to demonstrate its efficiency in supporting TDM traffic in OBS networks. The performance of the TCP traffic was also studied over a linear topology which showed that the throughput of the TCP degrades significantly beyond a BLP of 1% [103].

With the proposal of a technique to optically label the bursts based on an angle modulation scheme, the implementation of the LOBS was made possible in the year 2003 [82]. In the experiment, the optical label and the corresponding burst were modulated orthogonally to each other on the same wavelength using which the label was separated from the burst at the core nodes. By using the wavelength converters and electro-absorption modulators, label swapping was also demonstrated. The testbed developed at the BUPT was also enhanced to support labeled switching by including GMPLS control stack in the nodes developed earlier. This LOBS testbed was then used in several experiments to test the performance of the TCP as well as to implement grid computing applications. It was used to study the impact of various parameters in the OBS network on the instantaneous behavior of the TCP. Experimental results demonstrate that at a given BLP, the degradation in the throughput due to burst losses is much higher compared to that due to the loss of ACKs [102]. Due to the aggregation of packets and the burst losses in OBS network, the congestion window of TCP was observed to suffer from large fluctuations even during the transfer of large files. The fluctuations were minimized by reducing the offset time and the burst assembly time for the ACKs or by sending the ACKs in the control plane to avoid their losses [55]. Several high-speed TCP variants were also experimentally evaluated over the LOBS testbed to find out the one suitable for large file transfers in grid computing applications [42]. To enable the implementation of various applications over the OBS network, the LOBS testbed was developed to support end-user-initiated routing and switching. The testbed contained three edge nodes and four core nodes connected with bidirectional fiber links each with two data wavelengths and one control wavelength transmitting at 1.25 Gbps rate. Dynamic establishment of the end-to-end LSPs with the required QoS guarantees was experimentally demonstrated. The LSP setup messages are initiated as soon as the traffic arrives at the edge node. Using this testbed, the performance of a few of the prominent high-speed TCP variants was tested over the OBS network from which it was found that high-speed TCP (HSTCP) is the most suitable one for grid computing applications. As long as the BLP is low, TCP Westwood was found to have better performance than the others while fast TCP had poor performance at all the values of BLP.

As mentioned in the previous section the GGF and the researchers at the University of Essex have already made efforts to develop specifications for grid-over-OBS networks. Researchers at this university in collaboration with a few others across Europe have also experimentally demonstrated the feasibility of application-aware OBS networks. An ingress node interface capable of classifying grid traffic into appropriate bursts using GUNI interface is designed and demonstrated [99]. A grid-aware OBS network testbed with two edge nodes and a core node along with the GUNI and GRNI interfaces was also subsequently demonstrated [98]. To support a generic application in the future Internet over OBS networks, SIP is integrated with the OBS control plane and the proposed SIP-enabled OBS network is implemented in several experiments [12, 101]. A video-on-demand application was successfully demonstrated using the integration of SIP and JIT protocols in the control plane [100]. The SIP messages were encapsulated within JIT signaling messages which was shown to be very effective in managing application sessions to achieve one-step provisioning of the network resources.

To support grid applications in OBS networks, along with the SIP-based architecture, a peer-to-peer (P2P) architecture was also proposed and implemented over the LOBS testbed at the BUPT [40, 41]. The P2P architecture avoids the problems associated with the scalability and reliability of the centralized client–server architectures proposed earlier. However, due to the lack of standardization of the P2P messages and other problems associated with the security of the P2P systems, their implementation has some limitations. To combine the advantages of both P2P-based and SIP-based grid systems in OBS networks, a new architecture was proposed to obtain a hybrid SIP/P2P/OBS grid architecture [35]. An implementation of the hybrid architecture showed that it has the benefits of inter-operability, flexibility in session management, robustness, and high resource utilization. Experimental results showed that the SIP-enabled OBS network using the P2P architecture is more effective for future consumer grids compared to the other proposals.

1.8 Need for Modeling and Analysis of OBS Networks

The discussion so far in this chapter drives the point that there is a growing interest in the OBS technology around the world. Demonstration of the feasibility and applications of the OBS through testbed implementations in the recent times is believed to definitely create a flurry of activity in this field in the near future. The widespread activity of research both in the theory and in the practice of OBS networks provides the context for this book. Although not discussed in this book, the technology behind the devices used in OBS is also rapidly advancing which along with the testbed activities demands better understanding of these networks.

A key point to be remembered in the study of OBS networks is that they are designed to avoid processing and buffering in the optical domain at least till a viable technology is available for them. This is done by using a separate control plane to process the header information electronically while retaining the transmission of the data in optical domain. However, the main problem with this design is the possibility of burst losses due to contention at the intermediate nodes. The entire research in the OBS networks focuses on the techniques to minimize or eliminate the contention losses. Blocking probability of the bursts or the BLP thus becomes an important measure of the performance of these networks. This book focuses on this aspect of OBS networks and presents different analytical techniques developed to study the blocking probability in OBS networks.

Just like in any field of engineering, mathematical modeling and analysis is very essential to understand the behavior of the OBS networks and to develop efficient protocols. Mathematical modeling in OBS networks can be typically classified into two categories:

- Simple mathematical models are usually developed to obtain insights into the network phenomena. These models make a lot of simplifying assumptions to hide away many details of the network not directly related to the design of these networks. Such models are approximate in nature but flexible to be applied in different situations and often yield closed-form expressions enabling the interpretation of observable phenomena. For example, the blocking probability at a core node with full wavelength conversion is obtained by approximating it with an $M/M/k/k$ system used widely in the analysis of call blocking probability in the telecommunication networks. Although it ignores the effect of different parameters specific to OBS networks, the Erlang B formula for blocking probability is elegant enough to characterize the system in most of the simple cases.
- More advanced models that consider the effect of different parameters on the system are also developed but are evaluated using numerical techniques. Often

these models cannot be used to obtain any closed-form expressions but can be simplified using some heuristic techniques developed out of the simple models. The complex models are very effective in a detailed study of the effect of different network parameters on the system. For example, the blocking probability of OBS network is modeled through a system of fixed point equations which can only be numerically evaluated to obtain the operating point of the network. Such complex models are useful to evaluate the blocking probability in realistic network scenarios.

In the literature on OBS networks, there are many proposals to reduce the contention losses by efficient algorithms for routing and wavelength assignment, scheduling, signaling, and burst assembly. Some of the books on OBS networks published earlier discuss these techniques in great detail [17, 52, 85]. Most of the literature on OBS networks also focuses on techniques to mitigate the contention losses and support QoS at the higher layers. These proposals are independent of each other and are seldom compared against each other. Due to the existence of several approaches for reduction of the BLP, it is difficult to identify the best one for a given scenario only through simulation or experimentation. Although network practitioners and engineers argue that most of the technologies are typically developed with back-of-the-envelope calculations and precede any modeling exercise, this approach works only for simple scenarios. To extend the designs to any generic scenario requires a deeper understanding of their behavior. Analytical modeling is indispensable for this purpose because the generality of the conclusions is not possible with experimental investigation alone. For example, in the context of TCP such analytical models have provided very important results on the performance in different network technologies leading to development of different optimization techniques for each network. Therefore, the premise that quantitative modeling of networks has been and continues to be indispensable for the proper understanding of the networks (even in qualitative terms) is the motivation for taking this approach to the study of OBS networks.

As mentioned earlier, the main focus of the book is to understand the blocking behavior of OBS networks in different situations. To evaluate the blocking probability of OBS networks, it is important to understand the characteristics of the assembled traffic considering the input traffic and the burst assembly mechanisms. Since multiple packets are assembled into a burst, the burst traffic is expected to have different statistical characteristics from those of the input traffic. Different characteristics of the traffic in terms of the distribution of the burst length and the inter-arrival time will have different impacts on the BLP. Thus, it is essential to understand the traffic characteristics in OBS networks. Various techniques used to characterize the traffic in OBS networks considering the effect of the assembly mechanism and other modules at the edge node are presented in the next chapter. From the analysis of the traffic characteristics it can be seen that the burst arrival process can be assumed to be Poisson, despite the long-range dependence of the incoming traffic. Even though the burst size distribution turns out to be different for different burst assembly algorithms, the blocking probability can be determined fairly well with the Erlang B formula since all the distributions have finite first moment.

Evaluating the blocking probability of OBS networks is very important to understand the loss process in the network, compute the BLP in the network, and design efficient mechanisms to minimize the same. It is also challenging due to the lack of buffers at the core nodes and the bursty nature of the losses. The blocking probability changes with the degree of wavelength conversion at the core nodes, the offset time, and any other contention resolution mechanisms. Chapters 3 and 4 focus on different models to evaluate the blocking probability in various scenarios considering the impact of different network parameters. These models also help in understanding the benefit of using different contention resolution mechanisms in reducing the BLP. Such analysis can help to answer a question like does increasing the number of wavelengths give the same benefit as increasing the number of wavelength converters or the degree of conversion? It also helps to compare the benefit of different contention resolution mechanisms and to find out which mechanism is suitable for a given scenario. For example, analysis shows that there is instability in the network when deflection routing is used resulting in sudden fluctuations in the network throughput. This helps to develop some techniques to stabilize the network when deflection routing is used. Mathematical models for blocking probability when FDLs are used at the core nodes help in dimensioning the FDLs for a given value of BLP.

Finally, this book discusses the problems associated with the TCP over OBS networks and some techniques proposed to improve the performance of the TCP. Given that TCP is the widely used transport protocol and also likely to be adopted in the future optical networks, the study of the performance of different TCP implementations in OBS networks gained a lot of importance. Due to the inherent burst losses and the delay due to the burst assembly, traditional TCP variants were found to have poor performance in OBS networks. Mathematical models as well as experimental results revealed that the performance of TCP over OBS networks degrades drastically with BLP even as low as 1%. At the same time due to the correlated delivery of multiple segments in a single burst, the existence of an optimal burst assembly time that maximizes the throughput for a given BLP was also demonstrated. Such theoretical analysis of the performance of TCP helped in the design of different variants of the TCP specific to the OBS networks.

1.9 Organization of the Book

This chapter introduced the basic concepts of OBS networks along with the major issues of research. It also discussed different variants of OBS developed to reduce the contention losses and to provide guaranteed service to the traffic. Further, some important applications of the OBS and the testbed implementations carried out across the world were also briefly discussed.

Chapter 2 discusses models that characterize the traffic in OBS networks. The effect of the burst assembly algorithm and the waiting time at the ingress node on the input traffic is discussed with the help of various mathematical models. It also discusses the queueing network models for the ingress node developed to evaluate the distribution of the burst length, inter-arrival time, and the waiting time at the ingress node.

Chapter 3 discusses models used to evaluate the blocking probability of bursts in OBS networks. It starts with a discussion on simple models and then continues with more accurate models that evaluate the effect of offset time and the degree of wavelength conversion on the blocking probability. Finally, models for blocking probability in networks that support multiple classes of traffic are also discussed.

Chapter 4 presents analytical techniques that characterize the benefit of using various contention resolution mechanisms such as burst segmentation, wavelength preemption, deflection routing, and FDLs in reducing the blocking probability. It discusses in detail the models that evaluate the blocking probability when different variants of burst segmentation are used and then presents models to evaluate the impact of other contention resolution mechanisms as well.

Chapter 5 discusses the performance of different flavors of the TCP including those specifically developed for the OBS networks and presents models that evaluate the throughput of TCP considering the BLP and the increase in the RTT. It also discusses the effect of different parameters of the network on the throughput of the TCP and then presents some techniques used to improve the performance of the TCP.

References

1. Al Amin, A. et al.: Demonstration of deflection routing with layer−2 evaluation at 40Gb/s in a three-node optical burst switching testbed. IEEE Photonics Technology Letters 20(3), 178–180 (2008)
2. Al Amin, A. et al.: Development of an optical-burst switching node testbed and demonstration of multibit rate optical burst forwarding. Journal of Lightwave Technology 27, 3466–3475 (2009)
3. Akar, N., Karasan, E., Dogan, K.: Wavelength converter sharing in asynchronous optical packet/burst switching: An exact blocking analysis for Markovian arrivals. IEEE Journal on Selected Areas in Communications 24(12), 69–80 (2006)
4. Avo, R., Guerreiro, M., Correia, N.S.C., Medeiros, M.C.R.: A signaling architecture for consumer oriented grids based on optical burst switching. In: Proceedings of IEEE ICNS, p. 119 (2007)
5. Azim, M.M.A., Jiang, X., Ho, P.H., Horiguchi, S.: A new hybrid architecture for optical burst switching networks. In: Proceedings of HPCC, pp. 196–202 (2005)
6. Baldine, I. et al.: JumpStart deployments in ultra-high-performance optical networking testbeds. IEEE Communications Magazine 43(11), S18–S25 (2005)
7. Barakat, N., Sargent, E.H.: Dual-header optical burst switching: A new architecture for WDM burst-switched networks. In: Proceedings of IEEE INFOCOM, pp. 685–693 (2005)
8. Barakat, N., Sargent, E.H.: Separating resource reservations from service requests to improve the performance of optical burst-switching networks. IEEE Journal on Selected Areas in Communications 24(4), 95–107 (2006)
9. Blumenthal, D.J., Prucnal, P.R., Sauer, J.R.: Photonic packet switches: Architectures and experimental implementation. Proceedings of the IEEE 82(11), 1650–1667 (1994)
10. Callegati, F., Cerroni, W., Campi, A., Zervas, G., Nejabati, R., Simeonidou, D.: Application aware optical burst switching test-bed with SIP based session control. In: Proceedings of TridentCom, pp. 1–6 (2007)

11. Callegati, F. et al.: SIP-empowered optical networks for future IT services and applications. IEEE Communications Magazine **47**(5), 48–54 (2009)
12. Campi, A., Cerroni, W., Callegati, F., Zervas, G., Nejabati, R., Simeonidou, D.: SIP based OBS networks for grid computing. In: Proceedings of ONDM, pp. 117–126 (2007)
13. Cao, X., Li, J., Chen, Y., Qiao, C.: Assembling TCP/IP packets in optical burst switched networks. In: Proceedings of IEEE GLOBECOM, pp. 84–90 (2002)
14. Chen, Y., Turner, J.S., Pu-Fan, M.: Optimal burst scheduling in optical burst switched networks. Journal of Lightwave Technology **25**(8), 1883–1894 (2007)
15. Chlamtac, I. et. al.: CORD: Contention resolution by delay lines. IEEE Journal on Selected Areas in Communications **14**(5), 1014–1029 (1996)
16. Christodoulopoulos, K., Varvarigos, E., Vlachos, K.G.: A new burst assembly scheme based on the average packet delay and its performance for TCP traffic. Optical Switching and Networking **4**(2), 200–212 (2007)
17. Chua, K.C., Guruswamy, M., Liu, Y., Phung, M.H.: Quality of Service in Optical Burst Switched Networks. Springer, USA (2007)
18. Dolzer, K., Gauger, C., Spath, J., Bodamer, S.: Evaluation of reservation mechanisms for optical burst switching. AEU Journal of Electronics and Communications **55**(1) (2001)
19. Dueser, M., Bayvel, P.: Analysis of a dynamically wavelength-routed optical burst switched network architecture. Journal of Lightwave Technology **20**(4), 574–585 (2002)
20. Esteves, R., Abelem, A., Stanton, M.: Quality of service management in GMPLS-based grid OBS networks. In: Proceedings of ACM Symposium on Applied Computing, pp. 74–78 (2009)
21. Fan, Y., Wang, B.: Physical impairment aware scheduling in optical burst switched networks. Photonic Network Communications **18**(2), 244–254 (2009)
22. Ge, A., Callegati, F., Tamil, L.: On optical burst switching and self-similar traffic. IEEE Communications Letters **4**(3), 98–100 (2000)
23. Griffith, D., Sriram, K., Golmie, N.: Protection switching for optical bursts using segmentation and deflection routing. IEEE Communications Letters **9**(10), 930–932 (2005)
24. Guerreiro, M., Pavan, C., Barradas, A.L., Pinto, A.N., Medeiros, M.C.R.: Path selection strategy for consumer grid over OBS networks. In: Proceedings of IEEE ICTON, pp. 138–141 (2008)
25. Guo, H., Wu, J., Liu, X., Lin, J., Ji, Y.: Multi-QoS traffic transmission experiments on OBS network testbed. In: Proceedings of ECOC, pp. 601–602 (2005)
26. Guo, H. et al.: A testbed for optical burst switching network. In: Proceedings of OFC (2005)
27. Hsu, C.F., Liu, T.L., Huang, N.F.: Performance analysis of deflection routing in optical burst switched networks. In: Proceedings of IEEE INFOCOM, pp. 55–73 (2002)
28. Hsueh, Y.L. et al.: Traffic grooming on WDM rings using optical burst transport. Journal of Lightwave Technology **24**(1), 44–53 (2006)
29. Ito, T., Ishii, D., Okazaki, K., Sasase, I.: A scheduling algorithm for reducing unused time slots by considering head gap and tail gap in time sliced optical burst switched networks. IEICE Transactions on Communications **J88-B**(2), 242–252 (2006)
30. Jeong, M., Cankaya, H.C., Qiao, C.: On a new multicasting approach in optical burst switched networks. IEEE Communications Magazine **40**(11), 96–103 (2002)
31. Jeong, M., Qiao, C., Xiong, Y., Cankaya, H.C., Vandenhoute, M.: Tree-shared multicast in optical burst-switched WDM networks. Journal of Lightwave Technology **21**(1), 13–24 (2003)
32. Kim, J. et al.: Optical burst transport: A technology for the WDM metro ring networks. Journal of Lightwave Technology **25**(1), 93–102 (2007)
33. Kong, H., Phillips, C.: Prebooking reservation mechanism for next-generation optical networks. IEEE Journal of Selected Topics in Quantum Electronics **12**(4), 645–652 (2006)
34. Kozlovski, E., Dueser, M., Zapata, A., Bayvel, P.: Service differentiation in wavelength-routed optical burst-switched networks. In: Proceedings of OFC, pp. 774–775 (2002)
35. Liu, L. et al.: Experimental demonstration of SIP and P2P hybrid architectures for consumer grids on OBS testbed. In: Proceedings of OFC (2009)

36. Lee, G.M., Wydrowski, B., Zukerman, M., Choi, J.K., Foh, C.H.: Performance evaluation of optical hybrid switching system. In: Proceedings of IEEE GLOBECOM, pp. 2508–2512 (2003)
37. Li, J., Qiao, C., Chen, Y.: Recent progress in the scheduling algorithms in optical-burst-switched networks (invited). Journal of Optical Networking 3(4), 229–241 (2004)
38. Liang, O., Xiansi, T., Yajie, M., Zongkai, Y.: A framework to evaluate blocking performance of time-slotted optical burst switched networks. In: Proceedings of IEEE LCN, pp. 258–267 (2005)
39. Liu, J., Ansari, N., Ott, T.J.: FRR for latency reduction and QoS provisioning in OBS networks. IEEE Journal on Selected Areas in Communications 21(7), 1210–1219 (2003)
40. Liu, L., Hong, X., Wu, J., Lin, J.: Experimental demonstration of P2P-based optical grid on LOBS testbed. In: Proceedings of OFC (2008)
41. Liu, L., Hong, X., Wu, J., Lin, J.: Experimental investigation of a peer-to-peer-based architecture for emerging consumer grid applications. Journal of Optical Communications and Networking 1(1), 57–68 (2009)
42. Liu, L., Hong, X., Wu, J., Yin, Y., Cai, S., Lin, J.: Experimental comparison of high-speed transmission control protocols on a traffic-driven labeled optical burst switching network testbed for grid applications. Journal of Optical Networking 8(5), 491–503 (2009)
43. Liu, L., Yang, Y.: Fair scheduling in optical burst switching networks. In: Proceedings of ITC, pp. 166–178 (2007)
44. Liu, Y., Mohan, G., Chua, K.C.: A dynamic bandwidth reservation scheme for a collision-free time-slotted OBS network. In: Proceedings of IEEE/CreateNet Workshop on OBS, pp. 192–194 (2005)
45. Lu, K., Zhang, T., Jafari, A.: An anycast routing scheme for supporting emerging grid computing applications in OBS networks. In: Proceedings of IEEE ICC, pp. 2307–2312 (2007)
46. De Leenheer, M. et al.: An OBS-based grid architecture. In: Proceedings of IEEE GLOBECOM Workshop on High Performance Global Grid Networks, pp. 390–394 (2004)
47. De Leenheer, M. et al.: Anycast algorithms supporting optical burst switched grid networks. In: Proceedings of ICNS, p. 63 (2006)
48. De Leenheer, M. et al.: A view on enabling consumer oriented grids through optical burst switching. IEEE Communications Magazine 44(3), 124–131 (2006)
49. Maier, M.: Optical Switching Networks. Springer, USA (2008)
50. Masetti, F. et al.: Design and implementation of a multi-terabit optical burst/packet router prototype. In: Proceedings of OFC, pp. FD1.1–FD1.3 (2002)
51. Mazloom, A.R., Ghosh, P., Basu, K., Das, S.K.: Coconet: A collision-free container-based core optical network. Computer Networks 52(10), 2013–2032 (2008)
52. Mukherjee, B.: Optical WDM Networks. Springer, USA (2006)
53. Nakano, Y.: NEDO project on photonic network technologies - Development of an OBS node prototype and key devices. In: Proceedings of OFC (2007)
54. National LambdaRail: http://www.nlr.net
55. Nawaz, A., Hong, X.B., Wu, J., Lin, J.T.: Impact of OBS and TCP parameters on instantaneous behaviour of TCP congestion window on LOBS network testbed. In: Proceedings of OFC (2009)
56. Peng, S., Li, Z., He, Y., Xu, A.: TCP window-based flow-oriented dynamic assembly algorithm for OBS networks. Journal of Lightwave Technology 27(6), 670–678 (2009)
57. Qiao, C.: Labeled optical burst switching for IP-over-WDM integration. IEEE Communications Magazine 38(9), 104–114 (2000)
58. Ramamirtham, J., Turner, J.: Time-sliced optical burst switching. In: Proceedings of IEEE INFOCOM, pp. 2030–2038 (2003)
59. Ramaswami, R., Sivarajan, K.N.: Optical Networks: A Practical Perspective. Morgan Kaufmann, USA (1998)
60. Raza, M.A., Mahmood, W., Ali, A.: Hybrid control and reservation architecture for multidomain optical burst switched network. Journal of Lightwave Technology 26(14), 2013–2028 (2008)

61. Rosberg, Z., Zalesky, A., Vu, H.L., Zukerman, M.: Analysis of OBS networks with limited wavelength conversion. IEEE Transactions on Networking **14**(5), 1118–1127 (2006)
62. Rugsachart, A., Thompson, R.A.: An analysis of time-synchronized optical burst switching. In: Proceedings of IEEE HPSR (2006)
63. Sahara, A., Kasahara, R., Yamazaki, E., Aisawa, S., Koga, M.: The demonstration of congestion-controlled optical burst switching network utilizing two-way signaling - Field trial in JGN II testbed. In: Proceedings of OFC (2005)
64. Sahara, A. et al.: Demonstration of a connection-oriented optical burst switching network utilizing PLC and MEMS switches. Electronics Letters **40**(25), 1597–1599 (2004)
65. Sahara, A. et al.: Congestion-controlled optical burst switching network with connection guarantee: Design and demonstration. Journal of Lightwave Technology **26**(14), 2075–2086 (2008)
66. Sanghapi, J.N.T., Elbiaze, H., Zhani, M.F.: Adaptive burst assembly mechanism for OBS networks using control channel availability. In: Proceedings of IEEE ICTON, pp. 96–100 (2007)
67. Segarra, J., Sales, V., Prat, J.: An all-optical access-metro interface for hybrid WDM/TDM PON based on OBS. Journal of Lightwave Technology **25**(4), 1002–1016 (2007)
68. Shihada, B., Ho, P.H.: Transport control protocol in optical burst switched networks: Issues solutions, and challenges. IEEE Communications Surveys and Tutorials **Second Quarter**, 70–86 (2008)
69. Simeonidou, D. R. Nejabati (Eds.): Grid optical burst switched networks (GOBS) global grid forum draft (2005)
70. Siva Ram Murthy, C. Mohan, G.: WDM Optical Networks: Concepts, Design, and Algorithms. Prentice Hall PTR, USA (2002)
71. Sivaraman, V., Vishwanath, A.: Hierarchical time-sliced optical burst switching. Optical Switching and Networking **6**(1), 37–43 (2009)
72. Sun, Y. et al.: Design and implementation of an optical burst-switched network testbed. IEEE Communications Magazine **43**(11), S48–S55 (2005)
73. Tachibana, T.: Burst-cluster transmission with probabilistic pre-emption for reliable data transfer in high-performance OBS networks. Photonic Network Communications **17**(5), 245–254 (2009)
74. Tachibana, T., Kasahara, S.: Burst-cluster transmission: Service differentiation mechanism for immediate reservation in optical burst switching networks. IEEE Communications Magazine **44**(5), 46–55 (2006)
75. Tan, C.W., Mohan, G., Li, J.C.S.: Achieving multi-class service differentiation in WDM optical burst switching networks: A probabilistic preemptive burst segmentation scheme. IEEE Journal on Selected Areas in Communications **24**(12), 106–119 (2006)
76. Tanemura, T., A. Al Amin, Nakano, Y.: Multihop field trial of optical burst switching testbed with PLZT optical matrix switch. IEEE Photonics Technology Letters **21**(1), 42–44 (2009)
77. Teng, J., Rouskas, G.N.: Wavelength selection in OBS networks using traffic engineering and priority-based concepts. IEEE Journal on Selected Areas in Communications **23**(8), 1658–1669 (2005)
78. Tode, Y.H.H., Honda, H., Murakami, K.: Multicast design method using multiple shared-trees in OBS networks. In: Proceedings of OFC/NFOEC (2006)
79. Tode, Y.H.H., Murakami, K.: A novel cooperation method for routing and wavelength assignment in optical burst switched networks. IEICE Transactions on Communications **E90-B**(11), 3108–3116 (2007)
80. Venkatesh, T., Kiran, Y.V., Murthy, C.S.R.: Joint path and wavelength selection using q-learning in optical burst switching networks. In: Proceedings of IEEE ICC (2009)
81. Venkatesh, T., Sujatha, T.L., Murthy, C.S.R.: A novel burst assembly algorithm for optical burst switched networks based on learning automata. In: Proceedings of ONDM, pp. 368–377 (2007)

82. Vlachos, K.G., Monroy, I.T., Koonen, A.M.J., Peucheret, C., Jeppesen, P.: STOLAS: Switching technologies for optically labeled signals. IEEE Communications Magazine 41(11), S9–S15 (2003)
83. Vlachos, K.G., Ramantas, K.: A non-competing hybrid optical burst switch architecture for QoS differentiation. Optical Switching and Networking 5(4), 177–187 (2008)
84. Vokkarane, V.M., Jue, J.P.: Prioritized burst segmentation and composite burst-assembly techniques for QoS support in optical burst-switched networks. IEEE Journal on Selected Areas in Communications 21(7), 1198–1209 (2003)
85. Vokkarane, V.M., Jue, J.P.: Introduction to Optical Burst Switching. Springer, USA (2004)
86. Vokkarane, V.M., Jue, J.P.: Segmentation-based nonpreemptive channel scheduling algorithms for optical burst-switched networks. Journal of Lightwave Technology 23(10), 3125–3137 (2005)
87. Vokkarane, V.M., Zhang, Q.: Reliable optical burst switching for next-generation grid networks. In: Proceedings of BROADNETS, pp. 1428–1437 (2005)
88. Wang, X., Morikawa, H., Aoyama, T.: Burst optical deflection routing protocol for wavelength routing WDM networks. In: Proceedings of SPIE OptiComm, pp. 257–266 (2000)
89. Wei, J.Y., McFarland, R.I.: Just-in-time signaling for WDM optical burst switching networks. Journal of Lightwave Technology 18(12), 2019–2037 (2000)
90. Xin, C., Qiao, C., Ye, Y., Dixit, S.: A hybrid optical switching approach. In: Proceedings of IEEE GLOBECOM, pp. 3808–3812 (2003)
91. Xinwan, L., Jianping, C., Guiling, W., Hui, W., Ailun, Y.: An experimental study of an optical burst switching network based on wavelength-selective optical switches. IEEE Communications Magazine 43(5), S3–S10 (2005)
92. Xiong, Y., Vandenhoute, M., Cankaya, H.C.: Control architecture in optical burst-switched WDM networks. IEEE Journal on Selected Areas in Communications 18(10), 1838–1851 (2000)
93. Yang, L., Rouskas, G.N.: Adaptive path selection in OBS networks. Journal of Lightwave Technology 24(8), 3002–3011 (2006)
94. Yiannopoulos, K., Vlachos, K.G., Varvarigos, E.: Multiple-input-buffer and shared-buffer architectures for optical packet and burst switching networks. Journal of Lightwave Technology 25(6), 1379–1389 (2007)
95. Yin, Y., Hong, X., Wu, J., Kun, X., Yong, Z., Lin, J.: Experimental evaluation of the extended JET protocol in support of TDM services. In: Proceedings of OFC (2009)
96. Yoo, S.J.B.: Optical packet and burst switching technologies for the future photonic Internet. Journal of Lightwave Technology 24(12), 4468–4492 (2006)
97. Zalesky, A., Vu, H.L., Rosberg, Z., Wong, E.W.M., Zukerman, M.: Stabilizing deflection routing in optical burst switched networks. IEEE Journal on Selected Areas in Communications 25(6), 3–19 (2007)
98. Zervas, G., Nejabati, R., Simeonidou, D.: Grid-empowered optical burst switched network: Architecture, protocols and testbed. In: Proceedings of IEEE ICNS, p. 118 (2007)
99. Zervas, G., Nejabati, R., Simeonidou, D., M. O'Mahony: QoS-aware ingress optical grid user network interface: High-speed ingress OBS node design and implementation. In: Proceedings of OFC (2006)
100. Zervas, G. et al.: Demonstration of application layer service provisioning integrated on full-duplex optical burst switching network test-bed. In: Proceedings of OFC (2008)
101. Zervas, G. et al.: SIP-enabled optical burst switching architectures and protocols for application-aware optical networks. Computer Networks 52(10), 2065–2076 (2008)
102. Zhang, W., Wu, J., Lin, J., Minxue, W., Jindan, S.: TCP performance experiment on LOBS network testbed. In: Proceedings of ONDM, pp. 186–193 (2007)
103. Zhang, W., Wu, J., Xu, K., Lin, J.T.: TCP performance experiment on OBS network testbed. In: Proceedings of OFC (2006)
104. Zhang, Z., Liu, L., Yang, Y.: Slotted optical burst switching (SOBS) networks. Computer Communications 30(18), 3471–3479 (2007)

Chapter 2
Traffic Characteristics in OBS Networks

2.1 Introduction

Modeling and analysis of the OBS networks is a challenging task due to the unacknowledged switching used and the lack of optical buffers at the core nodes. Further, due to the aggregation of packets into bursts at the ingress nodes, the characteristics of the IP traffic are changed inside the OBS network. It is essential to analyze the characteristics of the traffic in the OBS networks to understand their performance. The characteristics of the traffic are mainly influenced by the burst assembly and disassembly as well as the burst loss in the core network. There are also a larger number of functional components in an ingress node compared to the egress node. Apart from this the ingress nodes are also responsible for the generation of BHPs and the determination of the offset time. Therefore, to study the characteristics of the traffic in the OBS networks, it is enough to analyze the components of an ingress node that are responsible for the formation and the transmission of the bursts. The function of the egress node is only to disassemble the bursts into the corresponding IP packets and forward them through the access network. Further, the traffic out of the egress node is not a part of the OBS network so that the influence of the egress node on the traffic characteristics can be neglected. The analysis of the traffic characteristics in the OBS networks helps to understand the performance of the network in terms of blocking probability, throughput, and delay.

At an ingress node, there are two main components that change the characteristics of the input traffic: the burst assembly mechanism and the burst assembly queue where the bursts await to be scheduled on an outgoing link. In addition to delaying the packets in the assembly queue, the edge node also changes the distribution of the inter-arrival time and size of the bursts depending on the assembly algorithm used. This chapter discusses the impact of the ingress node on the traffic characteristics in an OBS network with appropriate mathematical tools. The burst traffic is modeled using different input arrival processes and distributions of packet size. Some models are applicable only to specific input traffic distribution or packet size distribution while some of them are more generic. The effect of the burst assembly algorithm on the traffic characteristics is also analyzed in greater detail. Finally, this chapter presents some queueing network models for an ingress node to evaluate the throughput and the waiting time of the bursts at the ingress node.

T. Venkatesh, C. Siva Ram Murthy, *An Analytical Approach to Optical Burst
Switched Networks*, DOI 10.1007/978-1-4419-1510-8_2,
© Springer Science+Business Media, LLC 2010

2.2 Modeling Traffic at an Ingress Node

Analyzing the impact of ingress node on the traffic characteristics has two main benefits. It helps to understand the performance of the OBS networks for different distributions of burst size and inter-arrival time. It also helps to evaluate the end-to-end delay encountered by bursts in the OBS network. The main components to be considered in modeling the traffic in an OBS network are the arrival process of the IP traffic from the access network and the burst assembly mechanism. The burst formation time and the waiting time of the bursts in the transmission queue also increase the end-to-end delay of the packets. Burst aggregation at the ingress node also influences the distribution of the burst size, the inter-arrival time between the bursts, the burst formation time, and the number of packets in the burst. Similarly, the distribution of the input traffic at the ingress node influences the traffic in the core network. Some assumptions made about the arrival process of IP packets lead to a simpler description of the traffic. If the input traffic from the access network shows long-range dependence or self-similarity, the traffic characteristics are found to be different [9]. Simulation-based studies showed that the self-similarity in the input traffic is reduced by some extent due to the burst assembly mechanism which improves the throughput and the loss performance of OBS networks [5, 25]. However, theoretical studies showed that the degree of long-range dependence cannot be suppressed in general by the burst assembly, but the variation in the inter-arrival time between the bursts is reduced so that the burst assembly acts as a traffic shaping mechanism [8, 9, 27]. Therefore, considering the complex interaction of the components of the edge node with the input traffic, it is difficult to arrive at a generic model for the burst arrival process. The design of a generic model for the burst traffic is further complicated by the gamut of burst assembly mechanisms proposed in the literature. However, a few of the models discussed in this section serve as pointers to develop models for burst traffic with any general input arrival process.

2.2.1 Models for Specific Input Traffic Characteristics

In this section, the changes in the statistical properties of the assembled traffic when compared to the input traffic are studied with both short-range-dependent and long-range-dependent input traffic. The models for the assembled traffic presented in this section are applicable only to a particular input traffic and assembly mechanisms. For analytical tractability, some approximations are made based on the observations from simulations which make the final expressions easy to analyze. In the next section, models generic to any input traffic arrival process and packet size distribution are discussed which are also applicable to a wider range of burst assembly mechanisms. Before the traffic models are presented, a few of the terms used in this analysis are clarified through Fig. 2.1.

Assume that the IP packets arrive following a Poisson distribution with rate λ packets per second so that the inter-arrival time between the packets x_i is negative-exponentially distributed with parameter λ. Figure 2.1 shows the relation between

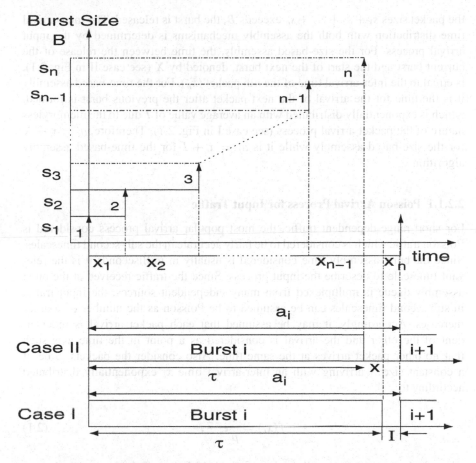

Fig. 2.1 Illustration of important terms used in the study of traffic characteristics

the packet sizes s_i, inter-arrival time between packets x_i, the inter-arrival time between bursts a_i, and the burst formation time τ, for both the time-based (case I) and the size-based assembly (case II) mechanisms. As shown in the figure, the packets, indicated by numbers $1, 2, \ldots, n - 1, n$ with sizes $s_1, s_2, \ldots, s_n$ arrive to form a burst. Depending on the trigger criteria used for the burst formation, the burst is released either when the size of the burst exceeds B or the timer T expires. The inter-arrival time between the bursts a_i is defined as the time between the start of the ith burst and the time when the next burst starts as indicated in the figure. For the time-based assembly the burst formation time denoted by τ is equal to the threshold value T as shown in case I. On the other hand, it is not constant for the size-based assembly mechanisms (case II). As shown in the figure, the burst formation time for the size-based assembly is determined by the input arrival rate which changes the number of packets required to fill up the threshold size B. As soon as the sum of the packet sizes $s_1 + s_2 + \ldots + s_n$ exceeds B, the burst is released. The inter-arrival time distribution with both the assembly mechanisms is determined by the input arrival process. For the size-based assembly, the time between the release of the current burst and the start of the next burst, denoted by X (see case II in Fig. 2.1), is equal to the inter-arrival time of the nth packet (x_n). For the time-based assembly it is the time for the arrival of the next packet after the previous burst is formed, which is exponentially distributed with an average value of I due to the memoryless nature of the packet arrival process (see case I in Fig. 2.1). Therefore, $a_i = \tau + X$ for the size-based assembly while it is $a_i = \tau + I$ for the time-based assembly algorithm.

2.2.1.1 Poisson Arrival Process for Input Traffic

For short-range-dependent traffic, the most popular arrival process considered is Poisson traffic which is considered to be fairly accurate in the sub-second timescales. Since the burst assembly time considered is usually in milliseconds it is the relevant timescale to describe the input process. Since the traffic received at the burst assembly queue is multiplexed from many independent sources, the input traffic in sub-second timescales can be assumed to be Poisson as the number of sources increases. Accordingly, it may be assumed that each packet arrival is independent of the other and the arrival is considered as a point in the time axis such that no other packet arrives at the same time. Also consider the packets to be of a constant size S arriving with an inter-arrival time x_i exponentially distributed according to

$$f(x_i) = \frac{1}{\mu} e^{-\frac{x_i}{\mu}}, \tag{2.1}$$

where μ is the mean as well as the standard deviation of the inter-arrival time. Assume that the time-based burst assembly algorithm is used with a constraint that the minimum burst length is not zero. In this case, the inter-arrival time between the assembled bursts is equal to the period of assembly T. Let the burst size be denoted by b_i which is essentially the sum of the sizes of n_i packets that arrive during T. The distribution of b_i can be evaluated from the basic principles of probability as

$$P[b_i = b] = P[n_i S = b]$$
$$= P\left[n_i = \frac{b}{S}\right]$$
$$= \frac{(\mu T)^{\frac{b}{S}-1}}{(\frac{b}{S}-1)!} e^{-\mu T}. \tag{2.2}$$

Similarly, if the size-based assembly algorithm is used with an upper bound on the time to wait before the burst is released, the burst size is always equal to the threshold value B. However, the inter-arrival time of the bursts denoted by a_i is a random variable which is essentially the sum of n_i independent inter-arrival times of the constituent packets. Note that $n_i = B/S$. Thus, the distribution of the inter-arrival time between the bursts can be evaluated as

$$
\begin{aligned}
P[a_i = a] &= P\left[\sum_{j=1}^{n_i} x_j = a\right] \\
&= P\left[\sum_{j=1}^{B/S} x_j = a\right] \\
&= \frac{(\mu b)^{\frac{B}{S}-1}}{(\frac{B}{S}-1)!} e^{-\mu b}.
\end{aligned}
\tag{2.3}
$$

The analysis presented above leads to some simple conclusions. The time-based assembly algorithm shapes the Poisson input traffic into traffic with a constant inter-arrival time (T) but with a variable burst size following a distribution given by Eq. 2.2 which tends to a mean value of ST/μ as the number of packets in a burst increases or the burst size increases. This can be easily obtained from the central limit theorem [20]. On the other hand the size-based assembly algorithm shapes the input traffic into traffic with constant burst size (B) but with an inter-arrival time distributed according to Eq. 2.3 which tends to a mean value of $\mu B/S$ from the central limit theorem. The distributions in Eqs. 2.2 and 2.3 tend toward Gaussian distribution asymptotically so that the assembled burst traffic either from the time-based or size-based algorithm becomes a constant rate traffic [26, 27].

Next, assume that the input traffic is still Poisson but the size of the packets is exponentially distributed with mean $1/S$, expressed as

$$
f(s_i) = \frac{1}{S} e^{\frac{-s_i}{S}}.
\tag{2.4}
$$

In case of the time-based assembly algorithm, the burst size distribution (given by Eq. 2.2 for constant packet size) gets modified to

$$
P[b_i = b] = P[n_i = n] f(s_i),
\tag{2.5}
$$

where the probability that there are n packet arrivals during T denoted by $P[n_i = n]$ is given by

$$
P[n_i = n] = \frac{(\mu T)^{n-1}}{(n-1)!} e^{-\mu T}.
\tag{2.6}
$$

Substituting Eqs. 2.4 and 2.6 in Eq. 2.5 gives the burst size distribution as

$$
P[b_i = b] = \frac{(\mu T)^{n-1}}{(n-1)!} e^{-\mu T} \frac{1}{S} e^{\frac{-s_i}{S}}.
\tag{2.7}
$$

For size-based assembly algorithm, the inter-arrival time between the bursts can be derived as

$$
P[a_i = a] = P\left[\sum_{j=1}^{n_i} x_j = a\right]
$$

$$
= \sum_{n=1}^{\infty} P\left[\sum_{j=1}^{n_i} x_j = a \mid n_i = n\right] P[n_i = n]
$$

$$
= \sum_{n=1}^{\infty} \frac{(\mu T)^{n-1}}{(n-1)!} e^{-\mu T} \times \frac{(SB)^{n-1}}{(n-1)!} e^{-SB}
$$

$$
= e^{-SB}\left[\sum_{n=0}^{\infty} \frac{(SB\mu)^n T^n}{n!}\right] e^{-\mu T} \tag{2.8}
$$

for which the lower and upper bounds can be obtained after some simplifications [26]

$$
\frac{1}{2} e^{-SB} e^{-\mu T + \sqrt{SB\mu T}} < P[a_i = a] < K e^{-SB} e^{-\mu T + \sqrt{SB\mu T}}, \tag{2.9}
$$

where $K > 1$ is a constant. From Eq. 2.9 it can be seen that for larger inter-arrival times the burst size distribution is similar to the exponential tail distribution due to $e^{-\mu T}$ being the dominating term. For the smaller inter-arrival times it is larger than the exponential distribution because the term $\sqrt{SB\mu T}$ dominates.

For Poisson input traffic, the assembled traffic tends to Gaussian distribution as the inter-arrival time becomes smaller leading to a smoother traffic. Since the burst inter-arrival time is nothing but a summation of the inter-arrival times of the constituent packets, the distribution of burst inter-arrival time depends on the distribution of the packet size. For the size-based assembly algorithm, the number of packets in a burst, nB/S also converges to the Gaussian distribution when the distribution of S converges to the Gaussian distribution. Similarly, the distribution of n approaches the Gaussian distribution from central limit theorem for larger values of n. However, if n is considered to be a fixed value because B is fixed, the distribution of the inter-arrival time approaches the Gaussian distribution and eventually converges to a constant. When the input traffic is Poisson with a variable packet size, the assembled traffic converges to the Gaussian distribution slowly compared to the case when the packet size is fixed, because there are two variables (packet size and packet inter-arrival time) to converge at the same time. The burst traffic from the time-based assembly algorithm also converges to the Gaussian distribution and becomes smooth with time for similar reasons.

2.2.1.2 Bernoulli Arrival Process for Input Traffic

Assume that the arrival process of the input traffic follows a slotted Bernoulli model where a single packet arrives in a given time slot with a probability p independent of the arrivals in the previous slots. The packet inter-arrival times x_i are geometrically distributed independent identically distributed (iid) random variables with mean $1/p$ and the probability generating function (pgf)

$$x(z) = E[z^{x_i}] = \frac{pz}{1 - (1 - p)z}. \tag{2.10}$$

The time slot is so small that there are no multiple arrivals in a slot. Also assume that the packet size is an iid random variable denoted by s_k with a probability mass function (pmf) $P[s_k = n]$. The associated pgf is given by

$$S(z) = E[z^{s_k}] = \sum_{n=1}^{S_{max}} z^n P[s_k = n], \tag{2.11}$$

where S_{max} is the maximum packet size. Let the distribution of the residual packet length, denoted by S_r be given by

$$P[S_r = n] = \frac{P[S > n]}{E[S]} \tag{2.12}$$

with the corresponding pgf given by

$$S_r(z) = \frac{1 - S(z)}{(1 - z)S'(1)}, \tag{2.13}$$

where $S'(z)$ is the first derivative of the pgf of the packet size. For most of the numerical calculations, the first two moments of S_r are sufficient which are related to the packet size S by

$$E[S_r] = \frac{E[S^2]}{2E[S]} - \frac{1}{2} \tag{2.14}$$

and

$$E[S_r^2] = \frac{E[S^3]}{3E[S]} - \frac{E[S^2]}{2E[S]} + \frac{1}{6}. \tag{2.15}$$

Therefore, computing the first three moments of the packet size is sufficient to analyze the burst traffic characteristics. Next, the burst traffic characteristics are analyzed for the above-mentioned packet size distribution and arrival process with the time-based and size-based assembly algorithms, and an algorithm with the thresholds on both time and size. Then using some approximations, simpler results are obtained which are used in the analysis of the traffic with different assembly algorithms.

Assume that for the time-based assembly algorithm, the threshold value is set to T time slots. Hence, each burst contains at least one packet that triggers the timer and all the packets that arrive in the subsequent T slots. Since it is assumed that there is at most one arrival in each time slot, the number of packets in a burst, n, can be obtained by adding the first packet and the number of packets during T slots. Therefore, n has the pgf

$$n(z) = (1 - p + pz)^T z. \tag{2.16}$$

Further, the size of the packet is assumed to be an iid random variable so that the burst size has the pgf given by

$$b(z) = n(S(z)) = S(z)(1 - p + pS(z))^T. \tag{2.17}$$

The first two moments of the burst size can be calculated from the above equation as

$$E[b] = E[S](1 + pT)$$
$$\text{Var}[b] = \text{Var}[S](1 + pT) + E[S]^2(1 - p)pT. \tag{2.18}$$

Note that the burst size distribution is obtained by taking the convolution of the packet size distribution with itself for a number of times that depend on the maximum packet size S_{max}. Numerically derived burst size distribution either by the convolution method or by taking the inverse transform of $b(z)$ hints at using a Gaussian distribution or Gamma distribution function as an approximation [11].

The burst size distribution is approximated by

$$P[b = B] \approx \frac{\alpha_b^{\beta_b} B^{\beta_b - 1}}{\Gamma(\beta_b)} e^{-\alpha_b B}, \tag{2.19}$$

where α_b and β_b are the parameters of Gamma distribution related to the first two moments of the burst size (defined by Eq. 2.18) by

$$\alpha_b = \frac{E[b]}{Var[b]}$$
$$\beta_b = \frac{E[b]^2}{Var[b]}. \tag{2.20}$$

Numerical results show that the Gamma distribution approximates the burst size distribution much closer than the Gaussian distribution especially around the tail ends of the distribution. Evaluating the Gamma distribution is also easy due to the availability of mathematical tables and numerical procedures [11].

Next, consider the assembly algorithm with a limit on the burst size, say B. Using the condition for the formation of a burst, the number of packets in a burst n satisfies the relation

$$n = i \quad \text{and} \quad b = B \quad \text{if and only if} \quad s_1 + s_2 + \cdots + s_{i-1} < B \leq s_1 + s_2 + \cdots + s_i = b.$$
$$(2.21)$$

Since the size of packets s_i is assumed to be an iid random variable, for $n \geq 1$ and $b \geq B$,

$$P[n = i, b = B] = P[s_1 + s_2 + \cdots + s_{i-1} < B \leq s_1 + s_2 + \cdots + s_n = b]$$

$$= \sum_{j=0}^{B-1} P[s_1 + s_2 + \cdots + s_{i-1} = j, B \leq j + s_i = b]$$

$$= \sum_{j=0}^{B-1} p^{(i-1)}(j) p(b - j), \qquad (2.22)$$

where $p^{(i-1)}(j)$ denotes the $(i - 1)$-fold convolution of the distribution $p(j) = P[s_i = j]$ which in turn can be expressed as the derivative of the $(i - 1)$-fold product of the pgf $s(x)$ given by

$$p^{(i-1)}(j) = \frac{1}{j!} \left(\frac{\partial^j}{\partial x^j} s(x)^{(j-1)} \right)_{x=0}. \qquad (2.23)$$

Taking a two-level z-transform of Eq. 2.22 with respect to both the variables and substituting the above expression for the convolution gives the joint pgf of n and b as

$$E[t^n z^B] = \sum_{i=1}^{\infty} \sum_{b=B}^{\infty} P[n = i, b = B] t^n z^B$$

$$= \sum_{i=1}^{\infty} \sum_{b=B}^{\infty} \sum_{j=0}^{B-1} \frac{1}{j!} \left(\frac{\partial^j}{\partial x^j} s(x)^{(i-1)} \right)_{x=0} p(b - j) t^n z^B$$

$$= \sum_{b=B}^{\infty} \sum_{j=0}^{B-1} \frac{1}{j!} \left(\frac{\partial^j}{\partial x^j} \frac{t}{1 - ts(x)} \right)_{x=0} p(b - j) z^B. \qquad (2.24)$$

For the polynomial $s(z)$ of degree k, the partial fraction expansion of

$$\frac{t}{1 - ts(x)} = \sum_{j=0}^{k-1} \frac{F_j(t)}{Q_j(t) - x} \qquad (2.25)$$

is used to calculate the partial derivatives at $x = 0$, where $Q_j(t)$ are the k roots of the denominator term $1 - ts(x) = 0$, and the function $F_j(t)$ is given by

$$F_j(t) = \lim_{x \to Q_j(t)} \frac{1}{1 - ts(x)} (Q_j(t) - x) = \frac{1}{s'(Q_j(t))}. \tag{2.26}$$

Using the partial fraction expansions and after some simplifications, Eq. 2.24 can be written as

$$E[t^n z^B] = \sum_{j=0}^{k-1} \frac{s(Q_j(t)) - s(z)}{s'(Q_j(t))(Q_j(t) - z)} \left(\frac{z}{Q_j(t)} \right)^B. \tag{2.27}$$

By using only one of the k functions in the expansion and also considering $Q_j(t) = 1$ for $t = 1$, the above equation can be further simplified to compute the approximate burst size distribution as

$$b(z) = z^B \frac{1 - s(z)}{s'(1)(1 - z)} = z^B S_r(z), \tag{2.28}$$

where $S_r(z)$ is the pgf of the residual packet size.

The approximate expression for the burst size can be understood by considering the burst assembly in the size-based assembly algorithm as a renewal process with lifetime s_i. That is, when the condition for the size-based assembly is satisfied, the burst size is given by

$$b = s_1 + s_2 + \cdots + s_i. \tag{2.29}$$

By the time the assembly reaches sufficiently large threshold B, it is near stochastic equilibrium so that the transient effects can be neglected. The amount by which the actual burst size b exceeds B is nearly given by the residual lifetime of the renewal process which is the same as the residual packet size S_r. Therefore the first two moments of the burst size can be approximated by

$$E[b] \approx B + E[S_r]$$
$$\text{Var}[b] \approx \text{Var}[S_r]. \tag{2.30}$$

Finally, consider an assembly algorithm where both the time and size thresholds are imposed simultaneously. To derive the burst size distribution in this case, it is enough to obtain the burst size distribution starting from the case $B = y$ to the case $B = \infty$ where the timer T expires. For this assembly algorithm any burst of size smaller than a threshold y should have been triggered by the expiry of T so that

$$P[b = x | B = y] = P[b = x | B = \infty] \quad \text{for} \quad x < y. \tag{2.31}$$

Conversely, for a burst of size bigger than the threshold y, the burst size threshold should have been exceeded in which case

$$P[b = x | B = y] \approx cP[S_r = x - y] \quad \text{for} \quad x \geq y, \tag{2.32}$$

where c is a constant which represents the probability that the burst is triggered on exceeding the size threshold given by

$$c = P[\text{size exceeds threshold} | B = y] = P[b \geq y | B = \infty]. \tag{2.33}$$

Approximating the burst size distribution with the Gamma distribution, c can be approximately written as

$$c \approx Z(\alpha_B(y - 0.5), \beta_B), \tag{2.34}$$

where $Z(\alpha, \beta)$ is an incomplete Gamma function

$$Z(\beta, x) = \frac{1}{\Gamma(\beta)} \int_x^\infty t^{\beta-1} e^{-t} dt. \tag{2.35}$$

The average burst size is thus given by

$$E[b | B = y] = \sum_{x=0}^{\infty} bP[b = x | B = y]$$

$$= \sum_{x=0}^{y-1} bP[b = x | B = y] + \sum_{b=y}^{\infty} bP[b = x | B = y] \tag{2.36}$$

which can be approximated as

$$E[b | B = y] = \sum_{x=0}^{y-1} bP[b = x | B = \infty] + c \sum_{b=y}^{\infty} bP[S_r = b - y]$$

$$\approx \int_0^{y-0.5} b \frac{\alpha_B^{\beta_B} b^{\beta_B-1}}{\Gamma(\beta_B)} e^{-\alpha_B b} db + c(y + E[S_r]). \tag{2.37}$$

Using some properties of the incomplete Gamma function, the above equation can be simplified to

$$E[b | B = y] \approx \frac{\beta_B}{\alpha_B}(1 - Z(\beta_B + 1, \alpha_B(y - 0.5))) + (y + E[S_r])Z(\beta_B, \alpha_B(y - 0.5)). \tag{2.38}$$

The second moment of the burst size can be evaluated in a similar fashion to be

$$E[b^2 | B = y] \approx \frac{\beta_B(\beta_B + 1)}{\alpha_B^2}(1 - Z(\beta_B + 2, \alpha_B(y - 0.5)))$$

$$+ (y^2 + 2yE[S_r] + E[S_r^2])Z(\beta_B, \alpha_B(y - 0.5)) \tag{2.39}$$

from which the variance of the burst size can be calculated. The above analysis shows that for low loads, the time threshold expires most of the time so that both the mean and the variance of the burst size show a linear dependence on p and hence the load. For high loads, the size threshold is the factor to trigger the burst so that almost all the bursts are of equal size $B + E[S_r]$ and the $\text{Var}[b] = 0$. For intermediate loads, the analysis presented above is inaccurate because of the approximations made. In the next section, the burst traffic is analyzed for a generic packet arrival process and with a generic size distribution. The analysis of burst traffic is also extended to more generic burst assembly mechanisms which help to evaluate the impact of burst assembly parameters on the traffic characteristics.

2.2.1.3 Exact Model for Poisson Traffic

In this section, the burst traffic generated by all the three popular burst assembly algorithms, namely time-based, size-based, and hybrid (time and size) assembly algorithms, is analyzed with Poisson-distributed input traffic. In the earlier section where the burst traffic is analyzed for Poisson input traffic, only time-based and size-based assembly algorithms are mainly considered. The hybrid algorithm had been approximated with the size-based algorithm with the assumption that the arrival rate is large enough to fill up the burst size before the timer expires. When the time-slotted Bernoulli process was considered for the input traffic, the distribution of the burst length is approximated by the Gamma distribution and the distribution of the inter-arrival time between the bursts is not considered. In this section, an exact model for Poisson arrival of packets with exponentially distributed sizes is presented. The assumption of exponentially distributed packet sizes simplifies the analysis and allows the synthesis of exact expressions for the hybrid burst assembly algorithm which otherwise would not be easy. It was validated with simulations that although the expressions are developed with the assumption of exponentially distributed packet sizes, they are almost accurate even for the trimodal distribution seen in real measurements [22, 23].

In realistic cases even if the burst assembly algorithm is purely time based or size based, there is a limit on the minimum burst length in the former and the maximum waiting time in the latter. Therefore, analyzing the hybrid algorithm which encompasses both the time-based and the size-based algorithms models the traffic for all the assembly mechanisms. In the case of time-based assembly algorithm with a threshold T, if there is no requirement on the minimum burst length, the probability distribution function (pdf) of the burst length $f_b(x)$ is given by

$$f_b(x) = \sum_{i=0}^{\infty} \frac{\mu(\mu x)^i}{i!} e^{-\mu x} \frac{(\lambda T)^i}{i!} e^{-\lambda T}, \quad (x > 0, \ i = 0, 1, \ldots), \qquad (2.40)$$

where λ is the arrival rate of the packets at the burst assembly queue and $1/\mu$ is the mean packet length, whereas the pdf of the inter-arrival time of the bursts is given by

$$f_a(x) = \lambda e^{-\lambda(x-T)}, \quad (x \geq T). \tag{2.41}$$

Note that since the packet arrival is Poisson the inter-arrival time between two bursts consists of a fixed part $\tau = T$ and a random value which is the time before the first packet in the subsequent burst arrives (indicated as X in Fig. 2.1). This is more exact than assuming that the timer is started only when the first packet arrives in a burst in which case the inter-arrival time between the bursts is assumed to be T. Note that $f_a(x)$ is independent of μ. While the effect of not considering X is negligible in heavy load conditions, it cannot be neglected in light load situations as it becomes comparable to the threshold value T.

If the restriction that the minimum burst length must be at least B_{min} is imposed for the time-based assembly, the burst is padded if the burst size $b < B_{min}$ which occurs with a probability

$$P_{pad} = P[b < B_{min}] = \int_0^{B_{min}} f_b(x)dx = \sum_{i=0}^{\infty} \frac{Z(i+1, \mu B_{min})}{i!} \frac{(\lambda T)^i}{i!} e^{-\lambda T}, \tag{2.42}$$

where $Z(x, y)$ is the incomplete Gamma function in Eq. 2.35. Hence, the pdf of the burst length with a possibility of padding is given by

$$f_{b,pad}(x) = f_b(x) + P_{pad}\delta(x - B_{min}), \quad (x \geq B_{min}), \tag{2.43}$$

where $\delta(.)$ is the Dirac delta function.

For the size-based assembly algorithm with a threshold B, the pdf of the inter-arrival time between the bursts is given by Eq. 2.8. In more realistic scenarios, there might be some delay between the departure of a burst and the arrival of the first packet in the subsequent burst. Assuming that the packets arrive following a Poisson process, similar to the distribution of the inter-arrival time of bursts in time-based assembly algorithm, the burst size is given by the shifted exponential distribution as

$$f_b(x) = \mu e^{-\mu(x-B)}, \quad (x \geq B). \tag{2.44}$$

Note that $f_b(x)$ is independent of the arrival rate of the packets λ. The shifted exponential distribution can also be interpreted by considering the fact that if the threshold B is actually exceeded after n packets arrive at the assemblers, the last packet that triggers the burst is not considered in the distribution of the burst size.

For the hybrid assembly algorithm, the exact expressions presented above are used to evaluate the distribution of the inter-arrival time and the burst size. Let the hybrid assembly algorithm be controlled by the size threshold B and the time threshold T apart from the requirement for the minimum burst length $B_{min} < B$. Assume that the timer is started when the first packet arrives at the assembly queue so that the probability that the timer expires before the size threshold is exceeded is given by

$$P_{TO} = 1 - P[\tau < T], \tag{2.45}$$

where τ is the random variable for the time required to satisfy the size threshold, i.e., the burst formation time, for which the pdf conditioned on the number of arrivals is given by

$$f_\tau(\tau|i) = \begin{cases} \delta(\tau), & \text{for } i = 0 \\ \frac{\lambda(\lambda\tau)^{i-1}}{(i-1)!}e^{-\lambda\tau}, & \text{for } i \geq 1. \end{cases} \tag{2.46}$$

Note that the number of arrivals i satisfies the constraint that the sum of the size of i packets equals B, i.e., $s_1 + s_2 + \cdots + s_i = B$ with the pdf of i arrivals given by Eq. 2.6. The case $i = 0$ indicates that the size of the first packet is greater than or equal to B and the timer is deactivated immediately after its activation so that $\tau = 0$. For the nontrivial case of $i \geq 1$, the pdf of τ is given by

$$P[\tau < T|i] = \int_0^T \frac{\lambda(\lambda\tau)^{i-1}}{(i-1)!}e^{-\lambda\tau}d\tau = \frac{Z(i, \lambda T)}{(i-1)!} \tag{2.47}$$

so that P_{TO} is given by

$$P_{TO} = 1 - e^{-\mu B} - \sum_{i=1}^{\infty} \frac{Z(i, \lambda T)}{(i-1)!} \frac{(\mu B)^i}{i!} e^{-\mu B}. \tag{2.48}$$

If the bursts leave the assembly queue before the timer expires, it definitely means that they are longer than B so that the pdf of the length of these bursts is similar to that in the size-based assembly algorithm given by Eq. 2.44. On the other hand if the bursts leave only after the timer expires, they have a length smaller than B which is distributed according to

$$f_{b,h}(x) = \begin{cases} P_{TO}f_b(x|x < B), & \text{for } x < B \\ (1 - P_{TO})\mu e^{-\mu(x-B)}, & \text{for } x \geq B \end{cases}, \tag{2.49}$$

where the conditional pdf of the burst length $f_b(x)$ given that $x < B$ is given by

$$f_b(x|x < B) = \frac{f_b(x)}{C} \tag{2.50}$$

with $f_b(x)$ defined in Eq. 2.40 and C being the normalization constant defined by [22]

$$C = P[b < B] = \sum_{i=0}^{\infty} \frac{Z(i+1, \mu B)}{i!} \frac{(\lambda T)^i}{i!} e^{-\lambda T}. \tag{2.51}$$

If the minimum burst length is not formed even after the timer expires, padding is done which occurs with a probability

$$P_{\text{pad},h} = P[b < B_{\min}] = \int_0^{B_{\min}} f_{b,h}(x)dx \quad \text{(for } B_{\min} < B)$$

$$= \int_0^{B_{\min}} P_{TO} f_b(x|x < B)dx = \frac{P_{TO}}{C} \left(\sum_{i=0}^{\infty} \frac{Z(i+1, \mu B_{\min})}{i!} \frac{(\lambda T)^i}{i!} e^{-\lambda T} \right). \tag{2.52}$$

Finally, the pdf of the burst length in a hybrid burst assembly algorithm, including the probability of padding, can be obtained from all the expressions derived so far as

$$f_{b,h,\text{pad}}(x) = \begin{cases} P_{\text{pad},h}\delta(x - B_{\min}) + P_{TO} f_b(x|x < B), & \text{for } B_{\min} \leq x < B \\ (1 - P_{TO})\mu e^{-\mu(x-B)}, & \text{for } x \geq B. \end{cases} \tag{2.53}$$

To compute the distribution of the inter-arrival time between the bursts, there are two parts to be considered, namely the burst formation time denoted by τ and the random part X which depends on the distribution of the inter-arrival time between the packets (refer Fig. 2.1). For the bursts that are released due to the expiry of the time threshold, $\tau = T$ so that the pdf of the inter-arrival time is given by Eq. 2.41. For all the other bursts, the pdf of the inter-arrival time, $f_a(.)$ can be obtained by conditioning on the case of $\tau < T$ so that

$$f_{a,h}(z) = P_{TO}\lambda e^{-\lambda(z-T)}U(z-T) + (1 - P_{TO})\mathscr{C}(z), \tag{2.54}$$

where $U(.)$ is the unit step function and $\mathscr{C}(z)$ is the convolution between the density functions of x and τ defined by

$$\mathscr{C}(z) = \lambda e^{-\lambda x} * f_\tau(\tau|\tau < T). \tag{2.55}$$

The conditional pdf of τ can be obtained from Eqs. 2.46 and 2.47 as

$$f_\tau(\tau|\tau < T) = \frac{1}{P[\tau < T]} \left(e^{-\mu B}\delta(\tau) + \sum_{i=1}^{\infty} \frac{\lambda(\lambda\tau)^{i-1}}{(i-1)!}e^{-\lambda\tau}\frac{(\mu B)^i}{i!}e^{-\mu B} \right), \tag{2.56}$$

where

$$P[\tau < T] = e^{-\mu B} + \sum_{i=1}^{\infty} \frac{Z(i, \lambda T)}{i - 1!} \frac{(\mu B)^i}{i!} e^{-\mu B}. \tag{2.57}$$

The convolution in Eq. 2.55 can be solved using Eq. 2.56 which in turn is used to evaluate the pdf of the inter-arrival time of the bursts. After some simplifications, the pdf of the inter-arrival time of the bursts for the hybrid assembly algorithm can be evaluated as

$$f_{a,h}(z) =$$
$$\begin{cases} P_{TO} \lambda e^{-\lambda(z-T)} + \frac{1-P_{TO}}{P[\tau<T]} \left(\lambda e^{-\mu B} e^{-\lambda z} + \sum_{i=1}^{\infty} \frac{\lambda(\lambda T)^i}{i!} e^{-\lambda z} \frac{(\mu B)^i}{i!} e^{-\mu B} \right), & z > T \\ \frac{1-P_{TO}}{P[\tau<T]} \left(\sum_{i=1}^{\infty} \frac{\lambda(\lambda z)^i}{i!} e^{-\lambda z} \frac{(\mu B)^i}{i!} e^{-\mu B} + \lambda e^{-\mu B} e^{-\lambda z} \right), & 0 < z \leq T. \end{cases}$$

To simplify this expression further, note that $B \gg \mu^{-1}$ in practice so that P_{TO} can be approximated by

$$P_{TO} \approx 1 - \sum_{i=1}^{\infty} \frac{Z(i, \lambda T)}{(i - 1)!} \frac{(\mu B)^i}{i!} e^{-\mu B} \tag{2.58}$$

using which the pdf of the inter-arrival time between the bursts can be approximated by

$$f_{a,h}(z) = \begin{cases} \left(1 - \sum_{i=1}^{\infty} \frac{Z(i,\lambda T)}{(i-1)!} \frac{(\mu B)^i}{i!} e^{-\mu B} \right) \lambda e^{-\lambda(z-T)} + \sum_{i=1}^{\infty} \frac{\lambda(\lambda T)^i}{i!} e^{-\lambda z} \frac{(\mu B)^i}{i!} e^{-\mu B}, & z > T \\ \sum_{i=1}^{\infty} \frac{\lambda(\lambda T)^i}{i!} e^{-\lambda z} \frac{(\mu B)^i}{i!} e^{-\mu B}, & z \leq T. \end{cases}$$

2.2.1.4 Self-Similar Input Traffic

If the traffic has long-range dependence, the results are slightly different for light and heavy load conditions. For the light load condition, the time-based assembly generates bursts of small size such that the processing time of the bursts is relatively smaller than their inter-arrival time. The burst assembly queue is usually empty in such cases due to which the traffic statistics are affected only during the assembly period. In large timescales of interest, the assembly time is negligible because of which the long-range dependency of the input traffic remains intact. On the other hand if the traffic load is heavy, the burst size is large and the burst assembly queue is not empty when a new burst is formed. Hence, the formation time of the previous burst affects the departure time of the current which increases the queueing and the processing time. The assembled burst traffic then tends to depart at a constant rate much similar to the packets queued at the bottleneck server in the Internet. However, due to the finite size of the buffer, this smoothing effect is only seen over smaller timescales while the long-range dependence is retained over larger timescales. In general the time-based assembly algorithm does not change the long-range dependency of the input traffic in the timescales of the order of the burst assembly time.

Assume that the traffic from the access network is modeled using a fractional Gaussian noise (FGN) process and the arrival rate from each source is independent of the other. Without loss of generality, assume that the burst is aggregated from several independent FGN processes so that the traffic at the relevant timescales is also described by the FGN process. Let the number of bytes that arrives from the source $1 \leq i \leq n$ in the time interval $(0, t]$ be denoted by $A_i(t)$ and the time threshold be T seconds. Due to the time stationarity of the distribution $\{A_i(t), t \geq 0\}$, the number of bytes per burst from a source i is equal to $A_i(T) - A_i(0)$ which is Gaussian distributed with mean μT and variance $\sigma^2 T^{2H}$, where μ, σ, and H are the mean, variance, and the Hurst parameter of the input FGN process [9]. Consider the timescales $t < T$, in which the number of burst arrivals denoted by $X(t)$ is approximately assumed to be Poisson distributed with parameter $\lambda = n/T$ with the first two moments being $t\lambda$ and $t\lambda(1 + t\lambda)$. On the other hand the total number of bytes in $(0, t]$, $A(t)$, has a Gaussian distribution since it is simply the sum of $X(t)$ iid Gaussian-distributed random variables $\{A_i(t)\}, 1 \leq i \leq n$. Therefore,

$$
\begin{aligned}
E[A(t)] &= E[E[A(t)|X(t)]] = E[A]E[X(t)] \\
E[A(t)^2] &= E[X(t)]\mathrm{Var}[A] + E[X(t)^2]E^2[A]
\end{aligned}
\tag{2.59}
$$

with $0 < t < T$ and zero otherwise. Thus the variance of the total number of bytes in $(0, t]$ is given by

$$
\mathrm{Var}[A(t)] = t\lambda(\mu^2 T^2 + \sigma^2 T^{2H})
\tag{2.60}
$$

and since

$$
\mathrm{Var}[A(t)/t] = \frac{\lambda}{t}(\mu^2 T^2 + \sigma^2 T^{2H})
\tag{2.61}
$$

the variance of the aggregated process $A(t)$ decays linearly with t. Therefore, $A(t)$ does not show long-range dependence for timescales smaller than T and the burst arrival process can be assumed to be Poisson.

However, this becomes inaccurate as $t \rightarrow T$, since the number of bursts is almost n. For timescales larger than T, the total input traffic at the assembler $A(t)$ can be written as $\sum_{i=1}^{n} A_i(t)$. Let $B_i(t)$ be the number of bytes from source i yet to be aggregated in a burst, which is Gaussian distributed with mean $\mu\delta t$ and variance $\sigma^2 \delta t^{2H}$ where δt is the time since the departure of the last burst. Then $A(t) = \sum_{i=1}^{n} A_i(t) - B_i(t) \approx \sum_{i=1}^{n} A_i(t)$ because the probability of packet arrivals beyond T decays exponentially for $t > T$ so that $B_i(t) \rightarrow 0$ as $t \rightarrow \infty$. In such a case, the variance of the OBS traffic converges to that of the input process which means that for larger timescales OBS traffic shows long-range dependence.

Similarly, the probability of no burst arrivals in the interval $[t, t + \delta t]$ denoted by $P[N_i(\delta t)]$ is given by the probability of no traffic from any source i during the time interval δt. This is equal to $A_i(\delta t) - A_i(0)$ which is a Gaussian-distributed random variable with mean $\mu \delta t$ and variance $\sigma^2 \delta t^{2H}$ so that

$$P[N_i(\delta t)] \approx e^{-\frac{\mu^2}{2(\delta t - T)^{2H} - 2\sigma^2}} \tag{2.62}$$

and the inter-arrival time between the bursts is distributed as

$$P[t > \delta t] \approx e^{-\frac{n\mu^2}{2(\delta t - T)^{2H} - 2\sigma^2}} . \tag{2.63}$$

With increasing number of sources n, the probability of inter-arrival times larger than T becomes negligible and the number of burst arrivals in smaller timescales can be approximated by a binomial distribution with parameters n and $\delta t / T$. At very small timescales such that $\delta t / T \rightarrow 0$ and large n, the number of arrivals can be approximated by a Poisson distribution with parameter λ and the inter-arrival time by a negative exponential distribution. As a consequence of these findings, there is nearly no influence of the self-similarity of the input traffic on the traffic in the OBS network in the timescales of the order of the time threshold values of the burst assembly mechanism. This means that the influence of self-similarity on the blocking probability can be safely neglected but it is useful in the dimensioning of optical buffers.

In the analysis so far, it is assumed that the burst is triggered by the expiry of the time threshold. Next, to study the worst-case delay performance at the edge node, assume that the burst is triggered when the size threshold exceeds. With either Poisson or self-similar input traffic, the worst-case queueing delay at the burst assembly queue can be understood only by considering large timescales for burst assembly. If the bursts assembled are larger, the variance in the inter-arrival times also becomes larger so that the delay performance at the transmission queue significantly depends on the variance of the aggregated traffic. To study this effect, let $P[W > w]$ denote the queueing delay distribution at the transmission queue which can be approximated by the complementary cumulative distribution function (ccdf) of the queue length at the transmission queue $P[X > x]$ approximated as $P[X > x] \approx P[W > x/C]$ where C is the capacity of the link serving the queue. For a single server queue, with Gaussian input traffic $A(t)$, the ccdf of the queue length is evaluated at the relevant timescale t which maximizes $\mathrm{Var}[A(t)]/(x + t(C - \lambda))$ where λ is the average rate of traffic [17]. This essentially means that the relevant timescale to study the self-similarity of the aggregated traffic is that which maximizes $P[X > x]$.

Assume that the IP traffic is approximated with a fluid flow model with $A(t)$ described by a Gaussian distribution with pdf of $f(t)$. Let the number of departures from the assembly queue in the time interval t be denoted by $D(t)$ with an inter-departure time of a. For larger timescales such that the size threshold B always dominates the burst assembly process, $E[D(t)] = E[A(t)]/B$. Let S_r be the residual burst size defined by $S_r = E[A(t)]/B - \lfloor E[A(t)]/B \rfloor$. The number of departures from the burst assembly queue in time t is then given by [7]

$$D(t) = \begin{cases} \lfloor E[A(t)]/B \rfloor + 1, & \text{with probability } S_r \\ \lfloor E[A(t)]/B \rfloor, & \text{with probability } 1 - S_r \end{cases} \quad (2.64)$$

with a variance given by

$$\begin{aligned} \text{Var}[D(t)] &= \int_0^\infty E[D(t)^2 | A(t) = t] f(t) dt - E[D(t)]^2 \\ &= \int_0^\infty \left((S_r(\lfloor \frac{t}{B} \rfloor + 1)^2) + (1 - S_r)(\lfloor \frac{t}{B} \rfloor)^2 \right) f(t) dt - \left(\frac{E[A(t)]}{B} \right)^2. \end{aligned}$$
$$(2.65)$$

Note that the above equation can be used to relate $D(t)$ with $A(t)$ as

$$\begin{aligned} \text{Var}[D(t)] &= \int_0^\infty \left((\lfloor \frac{t}{B} \rfloor + S_r)^2 + (S_r - S_r^2) \right) f(t) dt - \left(\frac{E[A(t)]}{B} \right)^2 \\ &= \int_0^\infty \left((\frac{t}{B})^2 + (S_r - S_r^2) \right) f(t) dt - \left(\frac{E[A(t)]}{B} \right)^2 \\ &= \int_0^\infty (S_r - S_r^2) f(t) dt + \frac{\text{Var}[A(t)]}{B^2}. \end{aligned} \quad (2.66)$$

Since it is difficult to solve the integral in Eq. 2.65 due to the floor function, it can be approximated for small and large values of t. For small timescales, i.e., $t \to 0$, $\text{Var}[A(t)]$ is small so that the pdf $f(t)$ tends to an impulse at λt and the term $\text{Var}[A(t)]/B^2$ can be ignored. Thus, for small timescales $\text{Var}[D(t)]$ can be approximated as

$$\text{Var}[D(t)] \approx (S_r - S_r^2)|_{\lambda t} = \frac{t}{E[a]} - \lfloor \frac{t}{E[a]} \rfloor - \left(\frac{t}{E[a]} - \lfloor \frac{t}{E[a]} \rfloor \right)^2 \quad (2.67)$$

which is actually the variance process of a constant bit-rate (CBR) flow with a constant inter-arrival time $E[a]$. This also shows that the assembled burst traffic resembles a CBR flow in small timescales.

On the other hand for large timescales, both $E[A(t)]$ and $\text{Var}[A(t)]$ are large and the span of pdf $f(t)$ is much larger than B. Therefore, by neglecting the effect of $f(t)$ in the first term of the integral in Eq. 2.66, the variance of the departure process can be approximately written as [7]

$$\text{Var}[D(t)] \approx \frac{1}{6} + \frac{\text{Var}[A(t)]}{B^2}. \quad (2.68)$$

This shows that in larger timescales, the variance process of the aggregated bursts is dominated by the variance of the input traffic so that the self-similarity of the input process is retained.

For smaller timescales, the variance of the aggregated traffic is similar to that of the CBR flow with arrival period equal to the mean burst departure time $E[D]$. The aggregated traffic can simply be described by the multiplexed traffic of n CBR flows. With this observation, the worst case delay at the burst transmission queue can be obtained by using the $nD/D/1$ model [20]. The $nD/D/1$ model describes a single queue serving n independent sources with the packets of equal size from all the sources. The worst case queueing delay in the transmission queue, W, can be obtained from the distribution $P[W > w]$ which can be approximated by the ccdf of the queue length, $P[X > x]$ for the $nD/D/1$ queue given by

$$P[X > x] = \sum_{x \leq i \leq n} \binom{n}{i} \left(\frac{i-x}{E[D]}\right)^i \left(1 - \frac{i-x}{E[D]}\right)^{n-i} \left(\frac{E[D]-n+x}{E[D]-i+x}\right). \quad (2.69)$$

Comparison of the theoretical results for the worst case waiting time in the transmission buffer with those obtained from the simulations show that the approximation of the transmission buffer with $nD/D/1$ queue is fairly accurate for small timescales when the traffic is smooth [7].

2.2.2 Models Independent of Input Traffic Characteristics

This section presents some theoretical models to characterize the traffic which do not make any assumptions on the packet size distribution or the input arrival process. While the analysis in the previous section uses different methodology for each distribution of the packet size and the input arrival process, the analysis presented here is more generic. Along with the distribution of the burst size and the inter-arrival time, the burst formation time and the number of packets per burst along with their moment generating functions are also evaluated. Using the moment generating function (mgf) makes it easy to compare different burst aggregation strategies for the mean and variance of the burst size, inter-arrival time, burst formation time, and the number of packets in each burst. The analysis presented here shows that the burst departure process from the traditional time-based and size-based burst assembly mechanisms is not purely Poisson which makes the computation of the blocking probability difficult. To alleviate this problem, a random burst assembly mechanism is designed to preserve the memoryless nature of the input traffic [21]. The approach presented here is also shown to be accurate when the input traffic is bursty which demonstrates that the analysis based on computing the mgf is applicable to a wider range of input traffic distributions and burst assembly mechanisms.

First, consider the time-based burst assembly mechanism in which the timer is triggered when the first packet arrives and the packet which arrives just after the timer expires is also included in the burst. Assume that the input traffic follows a

Poisson distribution with a parameter λ. Since the inter-arrival time between the IP packets is negative exponentially distributed with parameter λ, the inter-arrival time between the bursts a is also an exponential distribution shifted by T units with a pdf

$$f_a(t) = \lambda e^{-\lambda(t-T)}. \tag{2.70}$$

The mgf of the inter-arrival time between the bursts is given by

$$g_a(z) = e^{zT} \frac{\lambda}{\lambda - z} \tag{2.71}$$

from which the nth moment is calculated by evaluating the nth derivative of the mgf at $z = 0$. The average inter-arrival time between the bursts which is the first moment is given by

$$E[a] = \left(\frac{\partial g_a(z)}{\partial z} \right)_{z=0} = T + \frac{1}{\lambda}. \tag{2.72}$$

Since the timer is assumed to start with the arrival of the first packet in a burst, there is always one packet in the burst. Since one more packet is included in the burst after the timer expires, the mean inter-arrival time is greater than T. As illustrated in Fig. 2.1, the size of the burst at the end of time T is nothing but the sum of the size of the n packets that arrive in T seconds. The random variable n has a pmf given by

$$F_n(i) = \frac{(\lambda T)^{i-1} e^{-\lambda T}}{(i-1)!}, \quad \text{for } i = 1, 2, \ldots, \tag{2.73}$$

where $F_n(i)$ is the probability that there are i packets in a burst. The pdf of the burst size $b = s_1 + s_2 + \cdots + s_i$ is then given by

$$f_b(x) = \sum_{i=1}^{\infty} F_n(i) f_{s_i}(x), \tag{2.74}$$

where $f_{s_i}(x)$ is the convolution of the pdf of the packet size distribution $f_s(x)$ taken i times for i packets in a burst. Since it is not always possible to compute the convolution of the packet size distribution, the mgf of the burst size distribution is computed to simplify the computation of the burst size distribution for any general packet size distribution. Since b is a random sum of n iid variables which follow a Poisson distribution as seen from Eq. 2.73, the mgf is given by

$$g_b(z) = g_s(z).g_n(ln(g_s(z))) = g_s(z) e^{\lambda T(g_s(z)-1)}. \tag{2.75}$$

The mgf of the inter-arrival time between the bursts given by Eq. 2.71 and the mgf of the burst size given by Eq. 2.75 completely characterize the bursts in the case of the time-based assembly algorithm using which the mean and the variance can be computed for any general packet size distribution.

Second, for the size-based assembly algorithm the burst is released whenever the cumulative size of the packets accumulated in the buffer exceeds B. The number of IP packets in the burst can be computed from the probability that there are i packets in a burst conditioned on the fact that the cumulative size of i packets exceeds B but that of $i - 1$ packets does not. Therefore, the pmf of the number of packets in a burst is given by

$$
\begin{aligned}
F_n(i) &= \int_0^B P[S^i > B | S^{i-1} = b] f_{s_{n-1}}(b) db \\
&= \int_0^B [1 - X_s(B - b)] f_{s_{n-1}}(b) db, \quad \text{for } i \geq 1,
\end{aligned}
\tag{2.76}
$$

where $S^i = s_1 + s_2 + \cdots + s_i$ is the cumulative size of i packets and $X_s(.)$ is the cumulative distribution function (cdf) of the packet size s. The distribution of the number of packets can be used to evaluate the distribution of the inter-arrival time and the burst size for any general packet size distribution. The above equation gets simplified when a specific packet size distribution is substituted for $X_s(.)$ and it is also applicable to non-Poissonian arrival processes.

As illustrated in Fig. 2.1 the burst formation time τ for this assembly algorithm is a random variable described by $\tau = x_1 + x_2 + \cdots + x_{n-1}$ with n packets in a burst. With n characterized by the pmf in Eq. 2.76, the pdf of the burst formation time is given by

$$
f_\tau(t) = \sum_{i=1}^{\infty} P[\tau = t | n = i] F_n(i) = \sum_{i=1}^{\infty} f_{x_{i-1}}(t) F_n(i),
\tag{2.77}
$$

where $f_{x_{i-1}}(t)$ is an Erlang distribution of the cumulative packet size with parameters λ and $i - 1$. The inter-arrival time between the bursts is $a = \tau + E[x]$ for which the pdf is given by

$$
f_a(t) = \sum_{i=1}^{\infty} f_{x_i}(t) F_n(i)
\tag{2.78}
$$

and the mgf is similar to that in the case of time-based assembly given by

$$
g_a(z) = g_n(\ln(g_x(z))) = g_n\left(\ln\left(\frac{\lambda}{\lambda - z}\right)\right).
\tag{2.79}
$$

From Eq. 2.76, it can be seen that the number of packets in a burst is related to the packet size due to which the burst size cannot be simply expressed as the sum of the sizes of the IP packets. For the size-based assembly mechanism, since n

and s are not independent random variables, a joint pdf of b and n is evaluated by conditioning the probability that $b = S^n = b_1$ on $n = i$ as

$$f_{b,n}(b_1, i) = \int_0^B P[S^i = b_1|S^{i-1} = b_2]P[S^{i-1} = b_2]db_2$$

$$= \int_0^B f_s(b_1 - b_2)f_{S^{i-1}}(b_2)db_2, \quad \text{for } b_1 > B \text{ and } i \geq 1, \quad (2.80)$$

where it is assumed that the burst size $b_2 < b_1$ after the arrival of $i - 1$ packets while $b_1 > B$ after the last packet. Computing the marginal distribution of the above equation gives the burst size distribution to be

$$f_b(b_1) = \sum_{i=1}^{\infty} \int_0^B f_s(b_1 - b_2) \cdot f_{S^{i-1}}(b_2)db_2, \quad \text{for } b_1 > B \quad (2.81)$$

and 0 otherwise.

If the assembly algorithm is based on an upper limit on the number of packets in a burst, which is ideal for a class-based aggregation with different buffers for different classes of traffic, the number of packets in the burst is always fixed. To obtain the distribution for the inter-arrival time between the bursts and the burst formation time, note that the n packets in a burst are assumed to arrive following a Poisson distribution with negative exponentially distributed inter-arrival times. Therefore, the burst formation time is simply given by an Erlang distribution with parameters λ and $n - 1$ similar to the burst size distribution in the case of size-based assembly algorithm. The inter-arrival time between the bursts for this case is also given by $a = T + E[x]$ which results in the Erlangian distribution with parameters λ and n given by

$$f_a(t) = \frac{\lambda^n t^{n-1} e^{-\lambda t}}{(n-1)!} \quad \text{for } t \geq 0, n = 1, 2, \ldots. \quad (2.82)$$

The burst size distribution depends on the distribution of the packet size and is simply given by the n-fold convolution of the packet size distribution, $f_s(x)$, and its mgf would be equal to the mgf of the packet size raised to the power n.

Since the burst traffic resulting out of all the above-mentioned assembly algorithms is not necessarily Poisson, the blocking probability computed by assuming a Poisson arrival of bursts would not be accurate. To retain the Poissonian nature of the input traffic in OBS network, a simple random selection-based assembly algorithm is proposed in [21]. If the input traffic is Poisson with a parameter λ and a packet is selected out of the stream randomly with a probability p and sent to the burst, the resulting low-rate process is also Poisson with parameter $p\lambda$. A simple Bernoulli random generator at the burst assembler decides to send a packet to the burst or not based on the value of p which also is equal to the inverse of the average number of packets per burst. Using the definition of the random selection-based burst assembly mechanism, the probability that the number of IP packets in a burst is i is given by

$$F_n(i) = p(1 - p)^{i-1} \qquad (2.83)$$

corresponding to a geometric distribution with parameter p. The pdf of the burst size can be expressed similar to that in Eq. 2.74 where $F_n(i)$ is geometrically distributed as defined above. The mgf of the burst size is the mgf of the sum of i iid random variables s_i given by

$$g_b(z) = g_n(\ln(g_s(z))) = \frac{pg_s(z)}{1 - (1 - p)g_s(z)}. \qquad (2.84)$$

Using the random selection property of the Poisson process, every packet with an associated random number 1 can be assumed to mark the beginning of a burst. Since the inter-arrival time between the packets is negative exponentially distributed, the inter-arrival time between the bursts is also negative exponentially distributed with a pdf

$$f_a(t) = p\lambda e^{-p\lambda t}. \qquad (2.85)$$

The analysis presented in this section can be extended on similar lines for any general burst assembly mechanism to study its impact on the traffic characteristics in OBS networks. Next, the mgf-based approach used here is applied to study the traffic characteristics when the input arrival process is bursty.

2.2.2.1 Bursty Input Traffic

The approach of using the mgf in evaluating the traffic characteristics avoids the laborious computation of the exact distribution of the number of packets in a burst and the convolution of the packet size distribution, required to compute the distribution of the burst size and the inter-arrival time (for example, compare Eqs. 2.74 with 2.75). The first three moments of the burst size obtained from the mgf can be used to approximately determine the pdf of the burst size simplifying the evaluation of the traffic characteristics for any burst assembly algorithm [19]. This approach was used in the previous section to analyze the traffic characteristics when the input traffic follows Poisson distribution. The same approach is used here to analyze the traffic aggregated with different assembly algorithms with bursty input traffic modeled by the interrupted Poisson process (IPP) [16]. An IPP captures the notion of burstiness which is an important feature of the traffic in high-speed networks. The nature of the burst traffic when the input traffic is bursty is significantly different from that when the input traffic is Poisson which is seen from the analysis presented in this section.

    An IPP is an ON/OFF process, where the ON and the OFF durations are exponentially distributed with a mean of $1/t_1$ and $1/t_2$, respectively. The average duration of the ON and OFF periods is given by $t_2/(t_1 + t_2)$ and $t_1/(t_1 + t_2)$, respectively.

During the ON period, the arrival is Poisson distributed with rate λ while during the OFF period there are no arrivals. IPP is very useful to describe the data/voice and video transmission over the Internet where there is an idle time between the transmission of chunks of the data. Assume that the packet size is exponentially distributed with mean $1/S$. An IPP is also characterized by the coefficient of variation of the inter-arrival time of the packets given by

$$c^2 = 1 + \frac{2\lambda t_1}{(t_1 + t_2)^2} \tag{2.86}$$

which describes the burstiness of the traffic depending on the duration of the ON and OFF periods.

First, assume that the time-based burst assembly mechanism is used where the packets that arrive from an IPP during the burst assembly period T are aggregated into a burst. Let $g_n(t)$ be the mgf of the number of packets in a burst, $g_s(t)$ be the mgf of the packet size, and $g_b(t)$ be the mgf of the number of bytes that arrive during T. Then,

$$g_b(t) = g_n(\ln(g_s(t))). \tag{2.87}$$

Assuming the general case when the aggregation period T not only includes the ON period but also may include the OFF period, the number of packets in a burst may be different for each burst. In such a case, the state of the IPP changes from i at $t = 0$ to j at t and the probability that there are n arrivals during $(0, t]$ given that there are no arrivals at $t = 0$ is given by

$$P_{ij}(t, n) = P[N = n, S = j | N = 0, S = i] \tag{2.88}$$

which satisfies the Chapman–Kolmogorov equations. The z-transform of $P(t, n)$ is given by

$$P(t, z) = e^{(Q-(1-z)L)t}, \tag{2.89}$$

where Q is the infinitesimal generator of the IPP and L is the matrix of arrival rates defined by

$$Q = \begin{bmatrix} -t_1 & t_1 \\ t_2 & -t_2 \end{bmatrix} \quad \text{and} \quad L = \begin{bmatrix} \lambda & 0 \\ 0 & 0 \end{bmatrix}. \tag{2.90}$$

From the property of any discrete random variable, which says that the mgf of the random variable is equal to its z-transform when $z = e^s$, the mgf of the number of packets that arrive in $(0, t]$ is given by

$$g_n(t, s) = e^{(Q-(1-e^s)L)t}. \tag{2.91}$$

Replacing s with $\ln(g_s(-s))$, where $g_s(s) = S/(S+s)$ is the Laplace transform of the exponentially distributed packet size, the mgf of the burst size can be written as

$$g_b(t, s) = e^{(Q-(1-\frac{S}{(S-s)})L)t}.\tag{2.92}$$

The first two moments of the number of bytes that arrive during the burst aggregation period T can be computed as

$$m_1 = \mathbf{e}^T \left[\frac{\partial}{\partial s} g_b(T, s) \right]_{s=0} \mathbf{e},\tag{2.93}$$

where $\mathbf{e}$ is a column vector of 1s. To compute the derivative of the mgf, the eigenvalue decomposition method can be used. Let D be the diagonal matrix of the eigenvalues of the exponent A in Eq. 2.92 where

$$A = (Q - (1 - \frac{S}{(S-s)})L)t\tag{2.94}$$

and V be the matrix of the eigenvectors of A. Then the eigenvalue decomposition for the mgf in Eq. 2.92 is given by

$$e^{At} = Ve^{Dt}V^{-1}\tag{2.95}$$

so that the derivative of the mgf in Eq. 2.93 is obtained by taking the derivative of Eq. 2.95 using the chain rule as

$$\left[\frac{\partial}{\partial s} g_b(T, s) \right]_{s=0} = \frac{\partial e^{AT}}{\partial s}$$

$$= \frac{\partial V}{\partial s} e^{DT} V^{-1} + V e^{DT} \frac{\partial V^{-1}}{\partial s} + T V e^{DT} \frac{\partial D}{\partial s} V^{-1}.\tag{2.96}$$

Substituting the above expression for the partial derivative in Eq. 2.93 gives the first moment of mgf of the burst size as

$$m_1 = \frac{\lambda T}{S} \frac{t_2}{(t_1 + t_2)}\tag{2.97}$$

which actually shows that the mean number of bytes in a burst is equal to the mean number of bytes that arrive during the ON period multiplied by the average duration of the ON period (which is $t_2/(t_1+t_2)$). The second moment can similarly be written as

$$m_2 = \mathbf{e}^T \left[\frac{\partial^2}{\partial s^2} g_b(T, s) \right]_{s=0} \mathbf{e}\tag{2.98}$$

which can be obtained by using the chain rule on the first derivative given in Eq. 2.96. The pdf of the total number of bytes in a burst at the end of T can be obtained by approximating the moments with those of the two-stage Coxian distribution, C_2 [19]. The first three moments m_1, m_2, and m_3 are set equal to the first three moments of C_2 with parameters, s_1, s_2, and a. The three-moment fit can be used if $3m_2^2 > 2m_1m_3$ and $c^2 > 1$, otherwise a two-moment fit is used if $3m_2^2 \leq 2m_1m_3$ or $0.5 < c^2 < 1$. The pdf of C_2 is given by

$$f(x) = (1 - a)s_1 e^{-s_1 x} + a \left(\frac{s_1 s_2}{s_2 - s_1} e^{-s_1 x} + \frac{s_1 s_2}{s_1 - s_2} e^{-s_2 x} \right) \qquad (2.99)$$

so that in the case of a three-moment fit [19]

$$s_1 = \frac{X + (X^2 - 4Y)^{1/2}}{2}, \quad s_2 = X - s_1, \quad \text{and} \quad a = s_1 m_1 - 1,$$

where

$$X = \frac{1}{m_1} + \frac{m_2 Y}{2m_1} \quad \text{and} \quad Y = \frac{6m_1 - 3(m_2/m_1)}{6m_2^2/4m_1 - m_3} \qquad (2.100)$$

and in the case of a two-moment fit

$$s_1 = 2/m_1, \quad s_2 = 1/m_1 c^2, \quad \text{and} \quad a = 1/2c^2. \qquad (2.101)$$

The two-stage Coxian approximation cannot be used if $c^2 < 0.5$ in which case the generalized Erlang $E_{k-1,k}$ approximation is used with $1/k \leq c^2 \leq 1/(k-1)$ in which after the first exponential phase the service either continues in the rest of the $k - 1$ stages with a probability p or ends with a probability $1 - p$. The probability p is given by [19]

$$p = \frac{1}{1 + c^2} \left[kc^2 - k(1 + c^2) - k^2 c^2 \right]^{1/2} \qquad (2.102)$$

and the service rate of all the $k - 1$ stages put together is given by $\mu = (k - p)\lambda$. The pdf of the approximated burst size distribution in this case is given by

$$f(x) = p\mu^{k-1} \frac{x^{k-2} e^{-\mu x}}{(k-2)!} + (1 - p)\mu^k e^{-\mu x} \frac{x^{k-1}}{(k-1)!}. \qquad (2.103)$$

Second, assume that the size-based burst assembly algorithm is used where the threshold size is B and the packet size is negative exponentially distributed with mean S. For the case of Poisson arrivals, the analysis done in the previous section provides an accurate measure to characterize the distribution of the burst size including the size of the extra packet that triggers the release of the burst. Even for

IPP arrivals the same expression for the pdf given by Eq. 2.44 can be used because the burst is released when the number of exponentially distributed packets overflow the buffer which is independent of the way in which these packets arrive. The main effect of the IPP arrivals is on the burst formation time which is analyzed here. There is a strong equivalence between the IPP and the hyper-exponential distribution which is used to obtain the pdf of the inter-arrival time in the case of IPP arrivals as [4, 16]

$$f(t) = \frac{\lambda - s_1}{(s_1 - s_2)} s_1 e^{-s_1 t} + \left(1 - \frac{\lambda - s_1}{(s_1 - s_2)}\right) s_2 e^{-s_2 t}, \tag{2.104}$$

where the parameters of the C_2 distribution are given by

$$s_1 = \left[2\left(\lambda + t_1 + t_2 + \sqrt{(\lambda + t_1 + t_2)^2 - 4\lambda t_1}\right)\right]^{-1}$$

$$\text{and} \quad s_2 = \left[2\left(\lambda + t_1 + t_2 - \sqrt{(\lambda + t_1 + t_2)^2 - 4\lambda t_1}\right)\right]^{-1}. \tag{2.105}$$

The burst formation time τ can be expressed as a difference between the inter-arrival time between the bursts, a_i, and the time since the last packet that triggers the burst arrives, X, as indicated in case II of Fig. 2.1. Considering this, the first moment of τ can be written as

$$E[\tau] = \frac{SB}{\lambda} \frac{t_1 + t_2}{t_2} + \frac{1}{\frac{p}{s_1} + \frac{1-p}{s_2}} \left(\frac{2p^2}{s_1^2} + \frac{2(1-p)^2}{s_2^2}\right) + \frac{p-1}{s_2^2} - \frac{p}{s_1^2}, \tag{2.106}$$

where λ is the rate of arrival of packets and $p = (\lambda - s_1)/(s_1 - s_2)$. The second moment of the burst formation time can be obtained from the first moment after some simplifications:

$$E[\tau^2] = SB\left(\frac{p(2-p)}{s_1^2} + \frac{1-p^2}{s_2^2} - \frac{2p(1-p)}{s_1 s_2}\right)$$

$$+ \frac{S^2 B^2 \left(\frac{p}{s_1} + \frac{1-p}{s_2}\right)^2 + 8\left(\frac{p}{s_1^3} + \frac{1-p}{s_2^3}\right)}{\frac{p}{s_1} + \frac{1-p}{s_2}}$$

$$- 2\frac{SB(t_1 + t_2)}{\lambda t_2} \frac{\left(\frac{p}{s_1^2} + \frac{1-p}{s_2^2}\right)}{\left(\frac{p}{s_1} + \frac{1-p}{s_2}\right)} - 4\frac{\left(\frac{p}{s_1^2} + \frac{1-p}{s_2^2}\right)^2}{\left(\frac{p}{s_1} + \frac{1-p}{s_2}\right)}. \tag{2.107}$$

Finally, assume that the assembly algorithm has also a constraint on the minimum burst size apart from the upper bounds on the time and the size. This prevents very small bursts being released into the network. If the burst size is less than a minimum value say $B_{\min}$, the remaining bytes are kept in the assembly queue to be sent with the next burst. Similarly, if the burst size is larger than $B_{\max}$ when the timer expires, the remaining bytes are kept in the queue to be sent with the next burst. In this assembly algorithm, note that the IPP arrivals affect the distribution of both the number of packets in a burst as well as the size of the burst. If the residual bytes that might arise due to the burst length being larger than B_{max} are not considered, the first two moments of the burst size are similar to those in the case of the time-based assembly algorithm presented earlier. However, if there are residual bytes at the end of T, the pdf of the burst size $g_x(t)$ is given by the convolution of the pdf of the burst size at the end of T, $g_b(t)$, with the pdf of the residual burst size, $g_r(t)$. Therefore, the mgf of the number of bytes is given by

$$g_x(t) = g_b(t)g_r(t). \qquad (2.108)$$

Assume that the residual bytes are distributed uniformly in the burst size interval $[0, B_{min})$ so that the mgf can be evaluated easily from the analysis for time-based assembly as

$$g_x(t, s) = \begin{cases} e^{(Q-(1-\frac{S}{(S-s)})L)t} \frac{e^{tB_{min}}-1}{tB_{min}}, & \text{if } s > 0 \\ e^{(Q-(1-\frac{S}{(S-s)})L)t}, & \text{if } s = 0. \end{cases} \qquad (2.109)$$

Differentiating the mgf, the first two moments are evaluated which can be approximated with the C_2 distribution or the $E_{k-1,k}$ distribution to arrive at the pdf of the number of bytes arrived in T with the residual bytes added to the next burst. The approximate distribution for the residual burst size that is derived in the previous section for the case of Poisson arrivals holds good for the IPP arrivals which gives the cdf for the residual burst size to be

$$F_r(x) = \sum_{k=0}^{\infty} \int_0^{B_{min}} f_b(kB_{max} + x)dx, \qquad (2.110)$$

where $f_b(x)$ is the distribution of the burst size at the end of the period T satisfying $B_{min} \leq B \leq B_{max}$.

The analysis presented above is applicable to any burst assembly algorithm in general. Although this analysis is approximate due to the Coxian approximation used to compute the moments, it is shown to be accurate for almost any level of burstiness in the input traffic characterized by the coefficient of variation of the IPP. Comparing the distributions for burst size obtained for the cases of Poisson and IPP arrivals, a few observations can be made on the effect of the input traffic on the characteristics of the aggregated traffic. In the case of time-based assembly algorithm, the arrivals are slotted with the duration of the slot equal to the burst aggregation period. The burst size in this case is affected by aggregation period and also the input process. The burst size can also be expressed as a convolution of exponentially distributed random variables which tends to have a normal distribution for larger timescales. In the case of size-based aggregation, it is interesting to note that the burst size is not affected by the input arrival process. The packet size distribution and the size threshold are the only parameters that affect the burst size. The burst inter-arrival time distribution on the other hand is mainly affected by the input arrival process. In the case of Poisson arrivals, it is a convolution of exponentially distributed random variables while for IPP arrivals it is a convolution of hyper-exponentially distributed random variables.

2.3 Queueing Network Model for an Ingress Node

The ingress node is connected to a number of clients which are the IP routers or LSRs from which it receives the traffic. As mentioned in Chapter 1, IP packets destined to the same egress node and those that require the same level of service are assembled into a burst. Figure 2.2 shows the architecture of the ingress node with

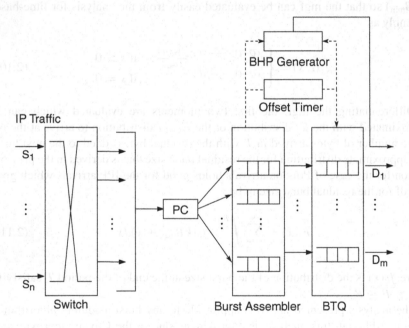

PC: Packet Classifier
BTQ: Burst Transmission Queue

Fig. 2.2 Queueing architecture of the ingress node

all the important modules analyzed in this section. Traffic that arrives at the ingress node from different IP routers (shown as S_1 to S_n) is classified by the classifier according to the destination node (shown as D_1 to D_m) and the class of service (if any). After the classification, packets are sent into different queues to be aggregated into bursts. For simplicity in implementation, bursts to different destination nodes and that belong to different classes of service are served by different burst assembly queues. Once the burst is assembled, a BHP is generated which carries all the information necessary for scheduling the burst at the core nodes. Depending on the number of hops to the destination and the processing time required at each node, the offset time is also determined for each burst. The bursts are queued in the burst transmission queue before they are scheduled on the outgoing link. Bursts destined to a particular egress node, which are usually served by the same outgoing link, are sent to the same burst transmission queue.

There are two main stages at the ingress node that need to be analyzed to understand the traffic characteristics of the OBS network. The first stage is the burst assembly queue that is responsible for the formation of the bursts. As seen in the previous section the characteristics of the traffic depends on the burst assembly algorithm and the input arrival process. In this section, the ingress node is analyzed as a network of different queues unlike the previous section which considers a single burst assembly queue. Since there could be multiple classes of traffic in the network, there are several burst assembly queues corresponding to the same egress node. Aggregation of traffic based on the classes allows the use of different criteria for burst assembly for each class of traffic and thus influences the burst arrival process into the network. In the next stage, the assembled bursts are queued in separate queues before they are transmitted on the outgoing link. The waiting time for the bursts in each queue may be different depending on the offered load on the outgoing link. At this stage there could also be prioritized scheduling to provide QoS guarantees. To evaluate the waiting time distribution of the bursts in this stage, the burst transmission queue should be modeled. It is usually assumed that the edge nodes have unlimited buffer capacity to store the bursts so that there is no loss of bursts at the edge nodes. Due to this assumption, the waiting time distribution of the bursts at the ingress nodes is often an important measure to be studied.

In this section, queueing theoretic models for the ingress node that help to analyze the above-mentioned parameters are considered. First, the buffer at the ingress node is assumed to be infinite so that after the bursts are formed they wait in the transmission queue till they are transmitted. The ingress node is modeled as a network of queues to evaluate the throughput of the node, i.e., the number of bursts served in a unit time. Next, the buffer at the ingress node is considered to be finite. The queueing network model for the ingress node in this case is used to evaluate the average number of bursts and the waiting time in the queue, throughput, and the blocking probability due to buffer overflow. Finally, models that consider the effect of burst assembly queue and the burst transmission queue on the waiting time of the bursts are presented.

2.3.1 Burst Arrival Process

To model the burst traffic at an ingress node, it is assumed that the traffic from the IP routers is aggregated at the ingress node which can be perceived as traffic into the network from some virtual sources [24]. In such cases the effect of the assembly algorithm on the traffic characteristics is not considered explicitly. Instead, the burst arrival process is assumed to follow a known distribution and the performance of the edge node is studied. One such model for the arrival process of the aggregated traffic is presented here which is used to study the waiting time and the throughput at the ingress node.

The arrival process of bursts on each wavelength is assumed to be independent of that on the other. The bursts arriving at the transmission queue are assumed to be from several independent ON/OFF sources with the ON and OFF times being independent and exponentially distributed [13, 24]. To conveniently model the variable length bursts when any generic burst assembly mechanism is used, a multistate Markov process is used to describe the arrival from each assembly queue. In general a wavelength is represented by different states when it is idle (no burst being transmitted) and when it is transmitting a burst. The state of the wavelength when it is transmitting a burst can be further split into multiple states depending on the length of the burst. This is the idea behind a multi-state Markov process used to describe the burst arrival process. Depending on the variation in the length of the bursts, the number of states can be increased to account for each distinct burst length.

A three-state Markov process presented here is used to characterize the state of the wavelength into idle, transmitting a short burst, or transmitting a long burst. Figure 2.3 shows the three-state Markov process that models the transmission of bursts with only two distinct lengths separated by an idle time. The mean duration of each of these states is characterized by the duration of short and long bursts as well as the idle time. The short bursts and the long bursts are exponentially distributed with lengths $1/t_s$ and $1/t_l$, respectively, and the mean duration of the idle state is $1/\phi$. The probability that a short burst arrives is given by p. Using these definitions it is easy to arrive at the transition rates of the three-state Markov process shown in the figure. If the length of the burst is not exponentially distributed, a Coxian distribution can be used to characterize the state of the wavelength. However, the analysis involving non-exponentially distributed burst lengths is quite involved so that the literature considers only Markov arrival process [24].

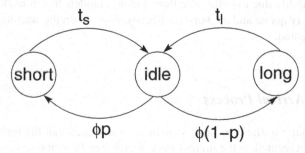

Fig. 2.3 ON/OFF burst arrival process

For the three-state Markov arrival process, the relationship between the inter-arrival time of bursts (τ_a), the burst duration (τ_l), and the idle time (τ_i) is given by [20]

$$L_{\tau_a}(s) = L_{\tau_l}(s)L_{\tau_i}(s) = \left(p\frac{t_s}{t_s + s} + (1 - p)\frac{t_l}{t_l + s}\right)\frac{\phi}{\phi + s}, \qquad (2.111)$$

where $L_{()}(s)$ is the Laplace transform of the corresponding variable. Differentiating the above equation gives the moments of the inter-arrival time as

$$E[\tau_a] = \frac{p}{t_s} + \frac{(1 - p)}{t_l} + \frac{1}{\phi}$$

$$E[\tau_a^2] = p\left(\frac{1}{t_s^2} + \frac{1}{\phi t_s} + \frac{1}{\phi^2}\right) + (1 - p)\left(\frac{1}{t_l^2} + \frac{1}{\phi t_l} + \frac{1}{\phi^2}\right), \qquad (2.112)$$

which describe the burst traffic. The burstiness of the arrival process which is measured by the squared coefficient of variation can be varied by adjusting the parameters of the three-state Markov process.

Since the bursts are either short or long, the service times are also described by either a short or a long duration. With p denoting the probability of a shorter service, the service time distribution is given by a two-state hyper-exponential distribution which is equivalent to a two-stage Coxian distribution with the squared coefficient of variation greater than or equal to unity. Let s_1 and s_2 denote the service rate of the first and second stages of the Coxian servers shown in Fig. 2.4, respectively, with p_c being the probability of transition from the first stage to the second.

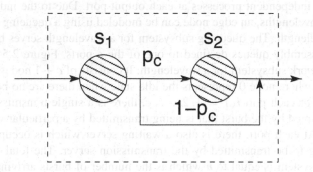

Fig. 2.4 Two-stage Coxian server

The parameters of the Coxian server are given by

$$s_1 = t_s, \quad s_2 = t_l, \quad \text{and} \quad p_c = \frac{(1 - p)(t_s - t_l)}{t_s}. \qquad (2.113)$$

The mean and the squared coefficient of variation for the burst duration are given by

$$E[\tau_l] = \frac{p}{t_s} + \frac{1-p}{t_l} \qquad\qquad (2.114)$$

$$CV^2[\tau_l] = \frac{2}{E^2[\tau_l]} \left(\frac{p}{t_s^2} + \frac{1-p}{t_l^2} \right) - 1. \qquad\qquad (2.115)$$

2.3.2 Modeling the Ingress Node with Infinite Buffer

In this section the ingress node is modeled using a network of queues with the input traffic characterized by the ON/OFF process described in the previous section. For simplicity, the traffic is assumed to arrive in the form of bursts ignoring the explicit characterization of the assembly algorithm. Each edge node is assumed to have c ports and the bursts are served by a link with w wavelengths. The buffer at the edge node is assumed to be infinite so that if a burst requests a wavelength that is busy it tries again after an exponential delay. Thus for each wavelength there may be many bursts waiting in the exponential back-off stage.

2.3.2.1 Node Without Wavelength Converters

If the core node does not have the wavelength conversion capability, the burst arriving on a wavelength i destined to a particular port can be switched only on the same output wavelength and the bursts on different wavelengths do not interfere with each other. The process of bursts awaiting service at an edge node can therefore be seen as w independent processes at each output port. Due to the independence among the wavelengths, an edge node can be modeled using a queueing subsystem for each wavelength. The queueing subsystem for a wavelength serves bursts from n different assembly queues destined to one of the c ports. Figure 2.5 shows the queueing network subsystem for a wavelength. It consists of $c + 1$ nodes numbered $0, 1, 2, \ldots, c$ where node 0 represents the idle state when there are no bursts being transmitted. For each port i, $i = 1, 2, \ldots, c$, there is a single transmission server which is occupied by the burst that is being transmitted by a particular wavelength at that port. At each port, there is also a waiting server which is occupied by the bursts waiting to be transmitted by the transmission server. The total number of bursts in the system is equal to n which is the number of bursts arriving from the assembler for transmission.

A burst arriving into the subsystem is initially in the idle state, i.e., node 0 for some time which may be assumed to be exponentially distributed with mean $1/\phi$. The burst then moves from node 0 to node i with a probability p_i where p_i denotes the probability that the burst has ith port as its destination. If the wavelength on which the burst arrived is free, the burst moves immediately to the transmission server, otherwise it moves to the waiting server and remains there for a duration

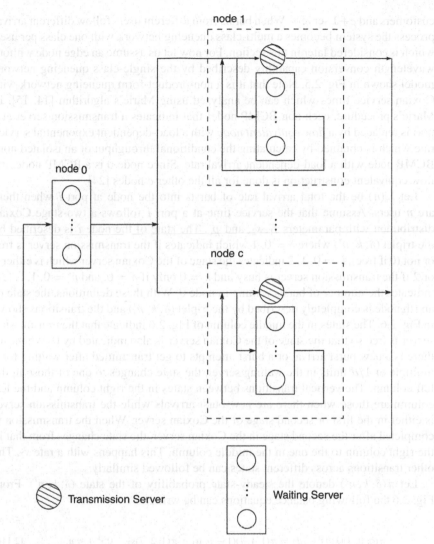

Fig. 2.5 Queueing network model of a subsystem at an edge node without wavelength converters

which is exponentially distributed with mean $1/d$. The burst requests the wavelength (transmission server) after the waiting time and remains in the waiting server till the wavelength is free. Bursts (users) in the waiting server are usually referred in the literature of queueing theory as *orbiting customers*. Once the transmission server is free the burst is served and the service time is exponentially distributed with a mean of either $1/t_s$ or $1/t_l$ with probabilities p or $1 - p$, depending on whether the burst is short or long, respectively.

Assume that the bursts from all the n clients have the same arrival process so that the closed queueing network model can be seen as a system with a single class of n

customers and $c+1$ servers. When bursts from different users follow different arrival process the system becomes a multi-class queueing network with one class per user, which is considered later in this section. For now let us assume an edge node without wavelength conversion capability described by the single-class queueing network model shown in Fig. 2.5. Note that it is a non-product-form queueing network with Coxian service times which can be analyzed using Marie's algorithm [14, 15]. In Marie's procedure, each non-BCMP node[1] that indicates a transmission server at a port is replaced by a *flow equivalent node* with a load-dependent exponential service rate which is obtained by calculating the conditional throughput of an isolated non-BCMP node with a load-dependent arrival rate. Since node 0 is a BCMP node, the flow equivalent construction is done for all the other c nodes [24].

Let $\lambda_i(n)$ be the total arrival rate of bursts into the node at port i when there are n users. Assume that the service time at a port i follows a two-stage Coxian distribution with parameters s_1, s_2, and p_c. The state of the node i is described by the triplet (δ, k, n') where $\delta = 0, 1$ which indicates if the transmission server is free or not (0 if free), $k = 0, 1, 2$ indicates the stage of the Coxian server which is either 1 or 2 if the transmission server is busy and $k = 0$ only if $\delta = 0$, and $n' = 0, 1, \ldots, n$ indicates the number of bursts waiting in node 0. With these definitions, the state of an ith node is completely described by the triplet (δ, k, n') and the transitions shown in Fig. 2.6. The states in the middle column of Fig. 2.6 indicate that the transmission server is free so that the stage of the Coxian server is also indicated by 0. Whenever there is a new burst arrival or a burst attempts to get transmitted after waiting for a multiple of $1/d$ units in the waiting server, the state changes to one of those in the left column. The vertical transitions between states in the right column and the left column are those when there are new burst arrivals while the transmission server is either in the first or second stage of the Coxian server. When the transmission is completed after the second stage in the Coxian server, the state changes from that in the right column to the one in the middle column. This happens with a rate s_2. The other transitions across different states can be followed similarly.

Let $\pi(\delta, k, n')$ denote the steady-state probability of the state (δ, k, n'). From Fig. 2.6 the following balance equations can be written:

$$\pi(0, 0, j)(\lambda(j) + jd) = \pi(1, 1, j)(1 - p_c)s_1 + \pi(1, 2, j)s_2, \quad 0 \le j < n, \tag{2.116}$$

$$\pi(1, 1, j - 1)(\lambda(j) + s_1) = \pi(0, 0, j)jd + \pi(1, 1, j - 2)\lambda(j - 1)$$
$$+ \pi(0, 0, j - 1)\lambda(j - 1), \quad 0 < j \le n, \tag{2.117}$$

$$\pi(1, 2, j - 1)(\lambda(j) + s_2) = \pi(1, 1, j - 1)p_c s_1 + \pi(1, 2, j - 2)\lambda(j - 1), \quad 0 < j \le n. \tag{2.118}$$

---

[1] BCMP network is a system of queues for which the steady-state occupancy of the queues has a simple joint-probability distribution or a product form. BCMP stands for Baskett, Chandy, Muntz, and Palacios, the authors of the seminal paper [1] that discusses this system of queues.

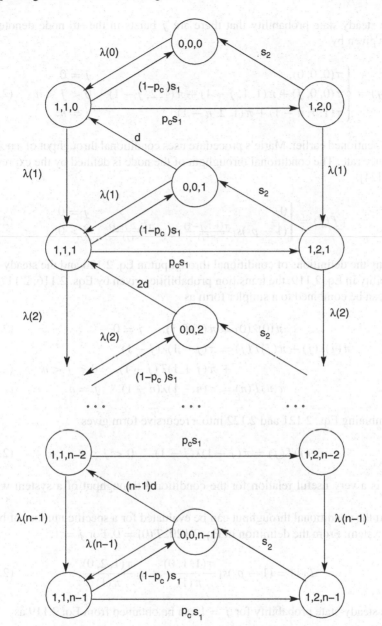

Fig. 2.6 State transition diagram for a node i in the subsystem of the queueing network model in Fig. 2.5

The steady-state probability that there are j bursts in the ith node denoted by $\pi(j)$ is given by

$$\pi(j) = \begin{cases} \pi(0, 0, 0), & j = 0 \\ \pi(0, 0, j) + \pi(1, 1, j - 1) + \pi(1, 2, j - 1), & 0 < j < n \\ \pi(1, 1, n - 1) + \pi(1, 2, n - 1), & j = n. \end{cases} \quad (2.119)$$

As mentioned earlier, Marie's procedure uses conditional throughput of a node as its service rate. The conditional throughput of the node is defined by the expression below [24]

$$T(j) = \begin{cases} 0, & j = 0 \\ (1 - p_c)s_1 \frac{\pi(1,1,j-1)}{\pi(j)} + s_2 \frac{\pi(1,2,j-1)}{\pi(j)}, & j > 0. \end{cases} \quad (2.120)$$

Using the definitions of conditional throughput in Eq. 2.120 and the steady-state probability in Eq. 2.119, the transition probabilities given by Eqs. 2.116, 2.117, and 2.118 can be combined to a simpler form as

$$\pi(0)\lambda(0) = \pi(1)T(1), \quad j = 0 \quad (2.121)$$

$$\pi(j)\lambda(j) + \pi(j)T(j) = \pi(j - 1)\lambda(j - 1)$$
$$+ \pi(j + 1)T(j + 1), \quad 0 < j < n \quad (2.122)$$
$$\pi(n)T(n) = \pi(n - 1)\lambda(n - 1), \quad j = n. \quad (2.123)$$

Combining Eqs. 2.121 and 2.122 into a recursive form gives

$$\pi(j)T(j) = \pi(j - 1)\lambda(j - 1), \quad 0 < j \le n, \quad (2.124)$$

which is a very useful relation for the conditional throughput of a system with j bursts.

Next the conditional throughput can be evaluated for a specific number of bursts in the system. From the definition in Eq. 2.120, $T(0) = 0$. For $j = 1$,

$$T(1) = (1 - p_c)s_1 \frac{\pi(1, 1, 0)}{\pi(1)} + s_2 \frac{\pi(1, 2, 0)}{\pi(1)}. \quad (2.125)$$

The steady-state probability for $j = 1$ can be obtained from Eq. 2.119 as

$$\pi(1) = \pi(0, 0, 1) + \pi(1, 1, 0) + \pi(1, 2, 0). \quad (2.126)$$

Similarly, from Eqs. 2.116 and 2.117,

$$\pi(0, 0, 1)(\lambda(1) + d) = \pi(1)\lambda(1) \quad (2.127)$$

$$\pi(1, 1, 0)(\lambda(1) + s_1) = \pi(0, 0, 1)d + \pi(1)T(1). \quad (2.128)$$

Solving Eqs. 2.125, 2.126, 2.127, and 2.128 simultaneously gives the conditional throughput of the system with one burst to be

$$T(1) = \frac{s_1 d(\lambda(1) - p_c\lambda(1) + s_2)}{(\lambda(1) + s_2 + p_c s_1)(\lambda(1) + d)}. \tag{2.129}$$

To obtain the conditional throughput for the system with any number of bursts, i.e., for $j > 1$, Eq. 2.116 can be rewritten using the definition of $T(j)$ as

$$\pi(0, 0, j)(\lambda(j) + jd) = \pi(1, 1, j)(1 - p_c)s_1 + \pi(1, 2, j)s_2$$
$$= \pi(j + 1)T(j + 1)$$
$$= \pi(j)\lambda(j). \tag{2.130}$$

Similarly, Eq. 2.117 can be rewritten as

$$\pi(1, 1, j - 1)(\lambda(j) + s_1)$$
$$= \pi(0, 0, j)jd + \pi(1, 1, j - 2)\lambda(j - 1) + \pi(0, 0, j - 1)\lambda(j - 1)$$
$$= \pi(0, 0, j)jd + \pi(j - 1)\lambda(j - 1)\frac{\pi(1, 1, j - 2) + \pi(0, 0, j - 1)}{\pi(j - 1)}$$
$$= \pi(0, 0, j)jd + \frac{\pi(1, 1, j - 2) + \pi(0, 0, j - 1)}{\pi(j - 1)}\pi(j)T(j) \text{ (from Eq. 2.124)}.$$
$$\tag{2.131}$$

From Eqs. 2.119, 2.120, 2.130, and 2.131 the following group of equations are obtained for the case of $j > 1$:

$$\pi(j) = \pi(0, 0, j) + \pi(1, 1, j - 1) + \pi(1, 2, j - 1)$$
$$\pi(j - 1) = \pi(0, 0, j - 1) + \pi(1, 1, j - 2) + \pi(1, 2, j - 2)$$
$$\pi(0, 0, j)(\lambda(j) + jd) = \pi(j)\lambda(j)$$
$$\pi(0, 0, j - 1)(\lambda(j - 1) + (j - 1)d) = \pi(0, 0, j)jd$$
$$\pi(1, 1, j - 1)(\lambda(j) + s_1) = \pi(0, 0, j)jd + \frac{\pi(1, 1, j - 2) + \pi(0, 0, j - 1)}{\pi(j - 1)}\pi(j)T(j).$$
$$\tag{2.132}$$

Solving the set of above equations for the unknown variables $\pi(0, 0, j)$, $\pi(0, 0, j - 1)$, $\pi(1, 1, j - 1)$, $\pi(1, 1, j - 2)$, $\pi(1, 2, j - 1)$, $\pi(1, 2, j - 2)$, and $T(j)$ gives the conditional throughput of the system with any $j > 1$ as

$$T(j) = \frac{js_1 d(\lambda(j - 1) + (j - 1)d)(\lambda(j) - p_c\lambda(j) + s_2)}{(\lambda(j) + jd)Z(j - 1)}, \tag{2.133}$$

where $Z(j - 1) = (j - 1)d(s_1 + s_2 + \lambda(j) - T(j - 1))$ $+ \lambda(j - 1)(p_c s_1 + s_2 + \lambda(j) - T(j - 1))$. The conditional throughput of the ith node

$T(i)$ is used as the load-dependent service rate of the ith node $\mu(i)$ in the Marie's algorithm given below.

Marie's algorithm is an iterative algorithm used to obtain the throughput of the non-closed-form queueing network of the type shown in Fig. 2.5. The main steps of the algorithm are as follows [14, 24]:

- *Step 1.* The service rate of the flow equivalent server $i = 1, 2, \ldots, c$ is set to $1/E[\tau_i]$ while it is set to ϕj for node 0 where j is the number of bursts in the waiting server.
- *Step 2.* For each node $i = 1, 2, \ldots, c$,

  - *Step 2.1.* The arrival rate of node i is calculated by short-circuiting node i in the substitute product-form closed queueing network, where each node i has an exponential service time of $\mu_i(j)$ for j customers.
  - *Step 2.2.* The conditional throughput of the node i is then computed using Eq. 2.129 or 2.133.
  - *Step 2.3.* The steady-state probability of node i is computed using Eq. 2.124.

- *Step 3.* The algorithm is terminated either if the sum of the mean number of bursts at all nodes is equal to the number of bursts in the queueing network or if the conditional throughput of each node is consistent with the topology of the queueing network; otherwise, step 2 is executed after setting the service rate equal to the conditional throughput for all the nodes.

2.3.2.2 Node with Wavelength Converters

If the node has full wavelength conversion capability, the above decomposition of an edge node into subsystems for each wavelength is not possible because the bursts arriving on different wavelengths might compete for the same wavelength on the outgoing link. In this case the edge node as a whole is modeled as a closed queueing network very similar to the case above. Figure 2.7 shows the queueing network model for an edge node with full wavelength conversion capability. In this model, there are $c + 1$ nodes corresponding to c ports and the idle state. Since the burst can be transmitted on any one of the w wavelengths, there are nw burst arrivals into the system. The burst initially enters node 0 and waits there till any one of the wavelengths at the destined port (node 1 through node c) is free. Since there is full wavelength conversion capability, there are w transmission servers at each port corresponding to w wavelengths on the outgoing link. If all the w transmission servers are busy, the burst enters the waiting server. As soon as a wavelength becomes free, the burst moves to one of the transmission servers where it is served for a duration which is exponentially distributed depending on the length of the burst. The remaining part is similar to the case where the node does not have wavelength conversion capability.

In this queueing network with multiple transmission servers in each node it is difficult to arrive at a closed-form solution for the conditional throughput of the node unlike the previous case. Instead, the queueing network for each node is solved numerically using the Gauss–Siedel method to get the steady-state probability that the node has j bursts. The conditional throughput is given by

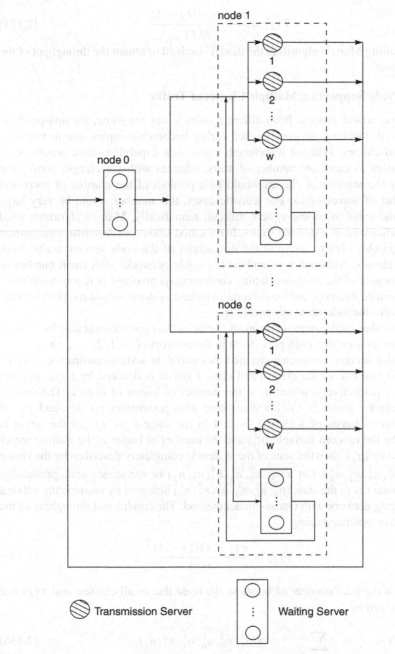

⊘ Transmission Server   ⋮ Waiting Server

Fig. 2.7 Queueing network model of a subsystem at an edge node with wavelength converters

$$T(j) = \frac{\pi(j-1)\lambda(j-1)}{\pi(j)} \tag{2.134}$$

and then using Marie's algorithm the model is solved to obtain the throughput of the ingress node.

2.3.2.3 Node Supporting Multiple Classes of Traffic

If the burst arrival process from different users is not the same, the non-product-form closed queueing network studied earlier becomes complex due to the large number of classes. Without wavelength conversion capability, there would be as many classes as there are number of users, whereas with wavelength conversion capability the number of classes would be a product of the number of users and the number of wavelengths. For realistic cases, this number would be very large making the solution of the system difficult numerically. Marie's algorithm used in the earlier case is extended to another method known as *heuristic aggregation method* (HAM) [18] to arrive at the throughput of the node serving traffic from multiple classes. Although it is applicable to only networks with small number of classes because of the time-consuming computations involved in it, any multi-class network can be decomposed into several two-class systems so that the HAM can be used to solve the individual systems [24].

Assume that there are two classes of bursts in the network indicated by 0 and 1 and further assume that each port (node in the network) $i = 1, 2, \ldots, c$ has $w \geq 1$ transmission servers. For each node, the flow equivalent node is constructed at which the arrival rate due to the class 0 and class 1 bursts is denoted by $\lambda_0(n_0, n_1)$ and $\lambda_1(n_0, n_1)$, respectively, where n_i is the number of bursts of class i. The service time of class j bursts is Coxian distributed with parameters s_{j1}, s_{j2}, and p_{cj}. If the number of bursts of a class $j = 0, 1$ in the stage $k = 1, 2$ of the server is denoted by the random variable $\{n_j^k\}$ and the number of bursts in the waiting server is denoted by $\{n'_j\}$, then the state of the system is completely described by the vector $(n_0^1, n_0^2, n'_0, n_1^1, n_1^2, n'_1)$. Let $\pi(n_0^1, n_0^2, n'_0, n_1^1, n_1^2, n'_1)$ be the steady-state probability that the node i is in the state $(n_0^1, n_0^2, n'_0, n_1^1, n_1^2, n'_1)$ obtained by numerically solving the queueing network with Gauss–Siedel method. The conditional throughput of the node is then evaluated using

$$T(j) = \frac{\pi(j-1)\lambda(j-1)}{\pi(j)}, \tag{2.135}$$

where j is the total number of bursts at the node due to all classes, and $\pi(j)$ and $\lambda(j)$ are given by

$$\pi(j) = \sum_{n_0^1+n_0^2+n'_0+n_1^1+n_1^2+n'_1=j} \pi(n_0^1, n_0^2, n'_0, n_1^1, n_1^2, n'_1), \tag{2.136}$$

$$\lambda(j) = \frac{1}{\pi(j)} \sum_{j_0+j_1=j} \pi(n_0^1, n_0^2, n'_0, n_1^1, n_1^2, n'_1) \times [\lambda_0(j_0, j_1) + \lambda_1(j_0, j_1)], \tag{2.137}$$

where $j_0 = n_0^1 + n_0^2 + n_0'$ and $j_1 = n_1^1 + n_1^2 + n_1'$. At each node, the conditional through-put is used as the load-dependent service rate of the flow equivalent node, and the convolution algorithm is used to obtain the arrival rates $\lambda_0(j_0, j_1)$ and $\lambda_1(j_0, j_1)$. Then Marie's algorithm is used till the convergence conditions are satisfied to solve the two-class queueing network. The HAM just described has limited applicability to networks with a small number of classes because it requires numerical analysis of a node with load-dependent arrivals for multiple classes and two-stage Coxian service time. Further, it also requires computation of the normalization constant. The complexity of HAM increases exponentially with the number of classes.

To solve the queueing network with a large number of classes, the original net-work is approximated with a set of two-class networks each of which is solved using the HAM [2, 24]. In this technique, for a $C = nw$ class network, C two-class networks are created. For each two-class network, the first class is class $i, i = 1, 2, \ldots, C$ in the original network while the second class is the aggregate class of all the other classes. To apply the HAM as in the previous two-class case, the Coxian server parameters for the aggregate class need to be determined at each node. To summarize the technique, let v_i be the visit ratio of class i which is the number of times a class is served. Assuming that the mean response time for class i is t_i, the throughput of class i bursts, T_i and the total throughput at the node T are related by

$$T_i = \frac{n_i}{t_c v_i}, \quad \text{and} \quad T = T_i v_i, \tag{2.138}$$

where n_i is the number of bursts of class i. The first two moments of the distribution of the aggregate class, a, i.e., the classes other than ith class, are given by

$$E[a] = \frac{\sum_{j \neq i} T_j E[\tau_l i]}{\sum_{j \neq i} T_j}$$

$$E[a^2] = \frac{\sum_{j \neq i} T_j E^2[\tau_l i](1 + CV^2(\tau_l i))}{\sum_{j \neq i} T_j}$$

$$CV^2[a] = \frac{E[a^2] - E^2[a]}{E^2[a]}, \tag{2.139}$$

where $E[\tau_l i]$ and $CV^2[\tau_l i]$ are the mean and squared coefficient of variation of the burst duration of class i defined in Eqs. 2.114 and 2.115, respectively. The service time distribution of the class a is approximated with the two-stage Coxian (C_2) distribution using moment matching technique for which the parameters are given by

$$s_{1a} = \frac{2}{E[a]}, \quad p_{ca} = \frac{1}{2CV^2[a]}, \quad \text{and} \quad s_{2a} = s_{1a} p_{ca}. \tag{2.140}$$

The detailed procedure to obtain the parameters for the aggregate class using the moment matching technique is given in [2, 15]. Once the service time distribution for both the classes is evaluated, the HAM is applied similar to the case when there are only two classes. In this way C two-class queueing network systems are solved simultaneously thereby reducing the complexity involved in solving the C-class system using the general HAM.

2.3.3 Modeling the Ingress Node with Finite Buffer

Consider the same architecture of an OBS network described in the previous section where each ingress node sends bursts from n sources to the other edge nodes. Assume that the bursts from an ingress node arrive following a Poisson process with exponentially distributed burst durations with a mean of $1/\mu$ seconds. The traffic from an ingress node is also assumed to be destined uniformly to all the other edge nodes and that the arrival rate from an ingress node is λ bursts per second. Unlike in the previous section where the buffer at the edge node is considered to be infinite, assume that a buffer of size $B > 1$ bursts is maintained for each destination node. Since the buffers are independent for each destination node, the arrival process from each buffer can be assumed to be independent of each other. Without loss of generality, assume that there is a single scheduler at the link serving all the buffers so that only one burst is transmitted at any time though there are idle wavelengths on the link.

With these assumptions, the edge node (referred to as the system henceforth) can be completely described by the n buffers since the bursts in different buffers are indistinguishable from those that are transmitted [29]. Let B_i be the number of bursts in the ith buffer ($1 \le i \le n$) including the one under service. The state of the system at time t is defined by

$$S(t) = (B_1, B_2, \ldots, B_n) \tag{2.141}$$

so that the total number of bursts in the system is $B_1 + B_2 + \ldots + B_n$. The number of states in the system is $(B + 1)^n$ which includes the case when all the buffers are empty. The state space is denoted by

$$\mathscr{S} = \{S(t) = (B_1, B_2, \ldots, B_n) | 0 \le B_i \le B, \ i = 1, 2, \ldots, n\} \tag{2.142}$$

which may be ordered in lexicographic ordering without loss of generality so that $(0, 0, \ldots, 0)_{n+1}$ is the first state and $(B, B, \ldots, B)_{n+1}$ is the last state.

To simplify the computation of state transition probabilities, consider only the state of the first buffer alone. This is enough since the evolution of the number of bursts in each buffer is assumed to be independent of that in the other. Since the arrival process is Poisson there is at most one arrival at the buffer $1 \le x \le n$ as well as only one departure from the buffer x in an infinitesimal interval of time Δt.

Let the state of the system at time t be denoted by $S_i = (B_1^i, B_2^i, \ldots, B_n^i)$ and the state at $t + \Delta t$ by $S_j = (B_1^j, B_2^j, \ldots, B_n^j)$. Since the buffers are independent of each other the transition between the elements of the state vector from t to $t + \Delta t$ occurs independent of the other. Let $P^{ij}(\Delta t)$ be the probability that the system changes its state from S_i to S_j and $P_x^{ij}(\Delta t)$ be the probability of transition for buffer x. From the assumption that the state of the buffers evolves independently

$$P_x^{ij}(\Delta t) = P[B_x^j | B_x^i]. \tag{2.143}$$

There are four possible cases in the computation of the conditional probability in Eq. 2.143 which are considered below.

Case 1. $B_x^j = 0, B_x^i \neq 0$: For this case, since there is only one departure from the buffer and no arrival in the interval Δt, $B_x^i = 1$. Let $T(B_1^i, B_2^i, \ldots, B_n^i)$ be the number of buffers that are busy at time t where a buffer is said to be busy if there is a burst being transmitted from the buffer at time t. If there are w wavelengths in the outgoing link, the busy buffers would contend for w wavelengths only if $T(B_1^i, B_2^i, \ldots, B_n^i) > w$. In this situation, a buffer x would win the contention with a probability of $w / T(B_1^i, B_2^i, \ldots, B_n^i)$. If $T(B_1^i, B_2^i, \ldots, B_n^i) < w$ then the buffer x wins the contention definitely. The departure rate from buffer x is given by

$$d(B_1^i, B_2^i, \ldots, B_n^i)\mu = \begin{cases} \dfrac{w\mu}{T(B_1^i, B_2^i, \ldots, B_n^i)}, & T(B_1^i, B_2^i, \ldots, B_n^i) > w \\ \mu, & T(B_1^i, B_2^i, \ldots, B_n^i) \leq w. \end{cases} \tag{2.144}$$

Therefore, the transition probability for the buffer x is

$$\begin{aligned} P_x^{ij}(\Delta t) &= (d(B_1^i, B_2^i, \ldots, B_n^i)\mu\Delta t + o(\Delta t))(1 - \lambda\Delta t + o(\Delta t)) \\ &= d(B_1^i, B_2^i, \ldots, B_n^i)\mu\Delta t + o(\Delta t). \end{aligned} \tag{2.145}$$

Case 2. $B_x^j = B_x^i = 0$: In this case, either there are no burst arrivals at the buffer x during Δt or there is a departure from the buffer during Δt. Therefore, the probability of transition from B_x^i to B_x^j is given by

$$\begin{aligned} P_x^{ij}(\Delta t) &= 1 - \lambda\Delta t + o(\Delta t) + (\lambda\Delta t + o(\Delta t))(d(B_1^i, B_2^i, \ldots, B_n^i)\mu\Delta t + o(\Delta t)) \\ &= 1 - \lambda\Delta t + o(\Delta t). \end{aligned} \tag{2.146}$$

Case 3. $B_x^j \neq 0$, $B_x^i = 0$: Similar to case 1, $B_x^j = 1$. There is a new arrival at the buffer but no departure during Δt. The transition probability B_x^i to B_x^j in this case is given by

$$\begin{aligned} P_x^{ij}(\Delta t) &= (\lambda\Delta t + o(\Delta t))(1 - d(B_1^i, B_2^i, \ldots, B_n^i)\mu\Delta t + o(\Delta t)) \\ &= \lambda\Delta t + o(\Delta t). \end{aligned} \tag{2.147}$$

Case 4. $B_x^j \neq 0$, $B_x^i \neq 0$: There are three events possible in this case which are mutually exclusive. First, $B_x^j = B_x^i + 1$ so that there is no departure from the buffer but there is a new burst arrival during Δt. The transition probability for this case is same as that given by Eq. 2.147. Second, there is no arrival but one departure from buffer x during the time interval Δt. Thus, $B_x^j = B_x^i - 1$ for which the transition probability is simply given by Eq. 2.145. Finally, there is neither an arrival nor a departure during Δt or there is an arrival and a departure from the buffer x. Thus, $B_x^j = B_x^i$ for which the probability of transition is given by

$$P_x^{ij}(\Delta t) = \begin{cases} 1 - \lambda\Delta t + o(\Delta t) - d(B_1^i, B_2^i, \ldots, B_n^i)\mu\Delta t, & \text{if } B_x^i \neq B \\ 1 - d(B_1^i, B_2^i, \ldots, B_n^i)\mu, & \text{if } B_x^i = B. \end{cases} \tag{2.148}$$

The transition probability matrix for the entire system is given by

$$\mathscr{P}(\Delta t) = [P^{ij}(\Delta t)]_{(B+1)^n \times (B+1)^n}, \tag{2.149}$$

where $P^{ij}(\Delta t)$ can be calculated by assuming that the buffers are mutually independent as

$$P^{ij}(\Delta t) = P_1^{ij}(\Delta t)P_2^{ij}(\Delta t)\ldots P_n^{ij}(\Delta t). \tag{2.150}$$

It is easy to see that the elements of $\mathscr{P}$ are zero iff $\sum_{x=1}^{n} |B_x^i - B_x^j| \geq 2$ so that there are $(B+1)^n + 2nB \times (B+1)^{n-1}$ non-zero entries. The total transition rate Q^{ij} can be obtained by

$$Q^{ij} = \begin{cases} \lim_{\Delta t \to 0} \frac{P^{ij}(\Delta t)}{\Delta t}, & \text{if } i \neq j \\ \lim_{\Delta t \to 0} \frac{P^{ii}(\Delta t)-1}{\Delta t}, & \text{if } i = j. \end{cases} \tag{2.151}$$

Let $\mathbf{Q} = [Q^{ij}]_{(B+1)^n + 2nB \times (B+1)^{n-1}}$ denote the transition rate matrix. The steady-state probabilities of the system denoted by vector $\pi = [\pi_0, \pi_1, \ldots, \pi_{(B+1)^n-1}]$ can be obtained using the balance equation of the Markov process along with the condition given below

$$\pi\mathbf{Q} = 0 \tag{2.152}$$

$$\sum_{i=0}^{(B+1)^n-1} \pi_i = 1. \tag{2.153}$$

To measure the performance of the system described above the following measures can be used:

• *Average number.* The mean number of bursts in the buffer x can be evaluated using the steady-state probabilities as

$$\overline{n}_x = \sum_{i=0}^{(B+1)^n-1} B_x^i \pi_i \qquad (2.154)$$

and the average number of bursts in the entire system is given by

$$\overline{n} = \sum_{x=1}^{n} \overline{n}_x. \qquad (2.155)$$

- *Throughput of buffer*. For a buffer x, the throughput is evaluated when the system reaches the steady-state condition when the number of bursts arriving into the system in a unit time interval equals the number of bursts departing from the system. The mean throughput of the buffer is given by

$$\tau_x = \lambda(B - \overline{n}_x). \qquad (2.156)$$

- *Average waiting time*. Using Little's theorem, the average delay in a buffer x can be obtained from the mean number of bursts and the throughput as

$$\partial_x = \frac{\overline{n}_x}{\tau_x}. \qquad (2.157)$$

The average number of bursts waiting for service in buffer x can be then written as

$$\overline{n}_x^w = \overline{n}_x - \sum_{i,B_x^i=0} \pi_i \qquad (2.158)$$

so that the mean waiting time in buffer x is

$$\partial_x^w = \frac{\overline{n}_x^w}{\tau_x} \qquad (2.159)$$

- *Utilization*. The utilization of the wavelengths in the system is defined as the ratio of the mean number of wavelengths that is busy to the total number of wavelengths which is also same as the ratio of the mean number of non-empty buffers to the number of wavelengths due to the assumptions made. The average number of non-empty buffers is given by

$$\sum_{i=0}^{(B+1)^n-1} T(B_1^i, B_2^i, \ldots, B_n^i)\pi_i = \sum_{x=1}^{n} \sum_{i=0}^{(B+1)^n-1} \text{sign}(B_x^i)$$

$$= \sum_{x=1}^{n} \sum_{i,B_x^i \neq 0} \pi_i = \sum_{x=1}^{n} (1 - \sum_{i,B_x^i=0} \pi_i). \qquad (2.160)$$

Therefore, utilization of the system is given by

$$u = \begin{cases} \dfrac{\sum_{x=1}^{n}(1-\sum_{i,B_x^i=0}\pi_i)}{w}, & \text{if } T(B_1^i, B_2^i, \ldots, B_n^i) < w \\ 1, & \text{otherwise.} \end{cases} \qquad (2.161)$$

- *Blocking probability*. A newly arriving burst at a buffer is blocked if it is full. When the system is in state $S_i = (B_1^i, B_2^i, \ldots, B_n^i)$, the number of buffers that is full would be

$$n_b = \sum_{x=1}^{n} 1 - \text{sign}(B - B_x^i). \qquad (2.162)$$

Assuming that a new burst arriving into the system has equal probability to select any one of the n buffers, the blocking probability of the system is given by

$$PB = \sum_{i=1}^{(B+1)^n-1} \frac{\sum_{x=1}^{n} 1 - \text{sign}(B - B_x^i)}{n}\pi_i. \qquad (2.163)$$

2.3.3.1 Approximate Solution

The model presented above is exact for an edge node with n buffers. It is easy to see that the complexity of the system increases exponentially with n due to the increase in the number of states ($(B + 1)^n$ in number). Hence, it is difficult to arrive at the exact solution presented above for a system with large number of buffers. If it is only enough to evaluate the average performance of the system, the exact model is not necessary. An approximate model for the system could be arrived at by not modeling the behavior of each of the buffers. Rather, the system state is modeled as the total number of bursts in the system ignoring the distribution of the bursts in the individual buffers. Further, ignoring the difference between the bursts waiting in the buffer and those being transmitted, the system model can ignore explicit consideration of the number of wavelengths.

In this case the state space of the system can be modeled as

$$\mathscr{S} = \{0, 1, 2, \ldots, i, \ldots, nB\}. \qquad (2.164)$$

Note that in the exact model, the number of bursts is given by $B_1 + B_2 + \ldots, +B_n$. By changing the representation of a state to $i = B_1 + B_2 + \ldots + B_n$, the system can be modeled as a single queueing network that approximately models the system. Assuming that the system is examined only at the steady-state condition, the dependence of the states on time is also neglected. With these simplifications, the system state $S = i$ is modeled as a continuous-time Markov chain (CTMC) with a finite number of states and Poisson arrival process at each buffer.

The approximate model just described is closely related to the restricted occupancy urn model whose properties can be used to evaluate the statistical behavior of the system [3, 29]. The restricted urn model is described by three variables, the balls, the urns, and a restriction on the number of balls in the urns. For the system of buffers considered here, note that the number of bursts and the number of buffers represent the balls and urns, respectively.

Let $A(i, n, B)$ be the number of ways in which i indistinguishable balls (bursts) are distributed among n distinguishable urns (buffers) under the B-restriction (buffer size limited to B), given by

$$A(i, n, B) = \sum_{x=0}^{n}(-1)^x \binom{n}{x}\binom{i+n-x(B+1)-1}{n-1}. \tag{2.165}$$

Let $A(i, n, B|p_t = u)$ be the number of ways in which i indistinguishable bursts are distributed among n distinguishable buffers such that u buffers have exactly t bursts under the B-restriction, given by

$$A(i, n, B|p_t = u) = \binom{n}{u}\sum_{x=0}^{n-u}(-1)^x\binom{n-u}{x} \times A(i-(u+x)t, n-(u+x), B). \tag{2.166}$$

Let $T_f(i)$ be the average number of full buffers in state i which can be evaluated using the urn model as

$$T_f(i) = \sum_{x=1}^{n} x\frac{A(i, n, B|p_B = x)}{A(i, n, B)}. \tag{2.167}$$

The arrival rate at state i (which is also the birth rate at state i) is given by

$$\lambda_i = \lambda(n - T_f(i)) \tag{2.168}$$

because new arrivals are accepted only if the buffers are not full. Let $T_b(i)$ be the average number of busy buffers in state i given by

$$T_b(i) = \sum_{x=1}^{\min(i,n)} x\frac{A(i, n, B|p_0 = n - x)}{A(i, n, B)}. \tag{2.169}$$

The death rate at state i is simply

$$\mu_i = \min(T_b(i), w)\mu, \tag{2.170}$$

where $\min(T_b(i), w)$ is used because the death rate is at most $w\mu$.

The steady-state probabilities can be computed from the equilibrium equations of the birth–death process [6]

$$\pi_i = \pi_0 \prod_{k=0}^{i-1} \frac{\lambda_k}{\mu_{k+1}}, \qquad (2.171)$$

where π_0 is given by

$$\pi_0 = \frac{1}{1 + \sum_{k=1}^{nB} \prod_{j=0}^{k-1} \frac{\lambda_j}{\mu_{j+1}}}. \qquad (2.172)$$

Once the steady-state probabilities are obtained the performance of the system is evaluated with the metrics below similar to the exact model.

- *System throughput.* The system throughput is evaluated from the theory of birth–death process as

$$\tau = \sum_{i=0}^{nB-1} \lambda_i \pi_i = \sum_{i=1}^{nB} \mu_i \pi_i. \qquad (2.173)$$

- *Average number.* The average number of bursts in the system is given by

$$\bar{n} = \sum_{i=0}^{nB} i \pi_i. \qquad (2.174)$$

- *Mean delay.* Mean delay in the system is then obtained using Little's theorem as

$$\partial = \frac{\bar{n}}{\tau} \qquad (2.175)$$

and the mean number of bursts waiting in the system excluding those being transmitted is

$$\bar{n}_w = \bar{n} - \tau. \qquad (2.176)$$

Therefore, the average waiting time in the system is

$$\partial_w = \frac{\bar{n}_w}{\tau}. \qquad (2.177)$$

- *Utilization.* The utilization of the channel is the ratio of the mean number of bursts successfully transmitted to the total number of wavelengths which is given by

$$u = \frac{\tau}{w}. \qquad (2.178)$$

- *Blocking probability.* The system is full when all the buffers are full. The blocking probability of the system is the probability that the buffer requested by the burst is full. Let $P(B|i)$ be the probability that a given buffer has B bursts when the system is in state i which is simply the probability that a given urn has exactly B balls given that i balls are distributed into n urns with at most B balls in each urn given by

$$P(B|i) = \frac{A(i - B, n - 1, B)}{A(i, n, B)}. \tag{2.179}$$

Therefore, the blocking probability of a buffer is given by

$$PB = \sum_{i=B}^{nB} P(B|i)\pi_i. \tag{2.180}$$

2.3.4 Modeling Waiting Time at the Ingress Node

The queueing network model of the ingress node in the previous section does not explicitly model the effect of the burst assembly algorithm. It assumes that the incoming traffic consists of bursts generated from a superposition of ON/OFF processes from several independent users. In such model, the waiting time of the bursts at the ingress node does not explicitly depend on the type of the burst assembly algorithm used. In this section, the waiting time of bursts in the burst assembly queue and also in the burst transmission queue is evaluated explicitly [12, 13]. It is assumed that the IP packets from the access network arrive at the ingress node which are then aggregated into bursts using a timer-based assembly algorithm. For simplicity it is assumed that the ON period of the traffic coincides with the burst assembly time so that all the packets generated during the ON period are assembled into a single burst. Using these assumptions, the mean number of packets and the average waiting time of packets in the burst assembly queue are evaluated. Once the burst is released after the expiry of the assembly time it waits in the burst transmission queue till it is transmitted on the outgoing wavelength. In this section, the data burst is assumed to be served in discrete time slots equivalent to integral multiples of the duration of the IP packets [13].

2.3.4.1 Modeling Burst Assembler Queue

Assume that the input arrival process at the burst assembly queue can be modeled as a superposition of n independent ON/OFF processes. Also assume that both the ON and OFF periods are geometrically distributed with the parameters a and b, respectively, so that the traffic to a burst assembly queue can be modeled as an interrupted Bernoulli process (IBP). The parameters a and b indicate the transition rates from OFF to ON and ON to OFF states, respectively. Let p be the probability

that an IP packet arrives from an active source during the ON period and all the
packets arriving during the ON period are aggregated into a single burst.

Let the state of the burst assembly queue be characterized by the number of IP
packets in it and π_i denote the steady-state probability that the burst assembly queue
contains i IP packets from a single ON/OFF process. The burst assembly queue can
then be modeled by the Markov chain shown in Fig. 2.8. A state i in the figure
indicates that there are i IP packets in the burst assembly queue.

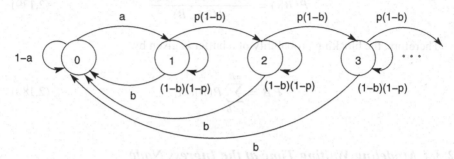

Fig. 2.8 Markov chain model for the burst assembly queue

The steady-state probability of state i denoted by π_i can be derived from the
transition diagram in Fig. 2.8 as

$$\pi_i = \begin{cases} \frac{b}{a+b}, & \text{for } i = 0 \\ \frac{a\pi_0}{p(1-b)+b}, & \text{for } i = 1 \\ \frac{p(1-b)\pi_{i-1}}{p(1-b)+b}, & \text{for } i \geq 2. \end{cases} \qquad (2.181)$$

The conditional probability that there are $i \geq 1$ IP packets in the burst assembly
queue is given by

$$\pi_{i|(i\geq1)} = \frac{\pi_i}{1-\pi_0}, \qquad i \geq 1 \qquad (2.182)$$

so that the steady-state probability in Eq. 2.183 can be written simply as

$$\pi_{i|(i\geq1)} = \begin{cases} \frac{b}{b+p(1-b)}, & \text{for } i = 1 \\ \left(1 - \frac{b}{b+p(1-b)}\right)\pi_{i-1|(i\geq1)} = \frac{b}{b+p(1-b)}\left(1 - \frac{b}{b+p(1-b)}\right)^{i-1}, & \text{for } i \geq 2, \end{cases}$$
$$(2.183)$$

which can be reduced further to

$$\pi_{i|(i\geq1)} = \frac{b}{b+p(1-b)}\left(1 - \frac{b}{b+p(1-b)}\right)^{i-1}, \qquad \text{for } i \geq 1. \qquad (2.184)$$

From Eq. 2.184 and the memoryless property of π_i, $\pi_{i|i \geq 1}$ has a geometric distribution so that the average number of IP packets in the burst assembly queue is given by

$$\overline{N} = \frac{b + (1-b)p}{b}. \tag{2.185}$$

The generating function for the random variable N, which represents the number of packets in the burst assembly queue, is given by

$$G_N(z) = \frac{\alpha}{1 - (1-\alpha)z}, \tag{2.186}$$

where $\alpha = b/(b + p(1-b))$.

Assume that the processing time required for each packet k during the generation of a burst, τ_k, is an independent and exponentially distributed random variable with mean $1/\mu$. The characteristic function for τ_k is given by

$$C_{\tau_k}(w) = \frac{\mu}{\mu - jw}. \tag{2.187}$$

Given that τ_k is the processing time for kth packet and the total time for the burst formation is

$$\tau = \sum_{k=1}^{N} \tau_k,$$

the mean burst formation time can be obtained from

$$E[\tau] = E[E[\tau|N]] = E[N.E[\tau_k]] = E[N]E[\tau_k]. \tag{2.188}$$

Therefore, the characteristic function of τ is given by

$$C_\tau(w) = E[E[e^{jw\tau}|N]] = E[F_{\tau_k}(w)^N]$$
$$= E[z^N]|_{z=C_{\tau_k}(w)} = G_N(C_{\tau_k}(w)). \tag{2.189}$$

Substituting Eqs. 2.186 and 2.187 into Eq. 2.189, the characteristic function of the burst formation time τ is evaluated as

$$C_\tau(w) = \frac{\alpha}{1 - \frac{\mu(1-\alpha)}{\mu - jw}}. \tag{2.190}$$

The inverse Fourier transform of Eq. 2.190 gives the probability density function of τ to be

$$f_\tau(x) = \alpha \left(1 + (1-\alpha)\mu e^{-\alpha\mu x}\right) \quad x \geq 0, \tag{2.191}$$

which signifies that the total time to assemble the burst from IP packets is also exponentially distributed. Therefore, the mean time to assemble a burst is given by

$$\bar{\tau} = \frac{1}{\alpha\mu} = \frac{b + p(1 - b)}{b\mu}. \tag{2.192}$$

2.3.4.2 Modeling Burst Transmission Queue

The burst transmission queue is modeled as a discrete-time system where the bursts are served in integral multiples of packets in each time slot. Without loss of generality, one IP packet is served in each time slot of duration equal to the packet duration. Assume that the length of the burst and the inter-arrival time are not correlated which is true if packets are assumed not to arrive back to back and the arrivals are not correlated. Note that from the definition of the IBP process assumed for the arrival, the average duration of the ON cycle is $1/b$ while it is $1/a$ for the idle time. The probability of a source being in the ON state is thus given by $P_{\text{ON}} = 1/bt$ where $t = 1/a + 1/b$ is the total duration of a cycle containing ON and OFF times. The probability of transition for a source from ON to OFF state is thus

$$P_s = P_{\text{ON}}b. \tag{2.193}$$

Let x_i denote the steady-state probability that the burst queue has i packets when the number of active sources is n so that the components of probability vector $\vec{x_i} = \{x_{i,j}\}$, $1 \le j \le n$ in general represent the number of packets in a burst queue with j active sources out of n in any time slot. The aggregate arrival process to the burst queue from n independent ON/OFF sources can be described by a discrete-time batch Markov arrival process (DBMAP). Since the state of the queue is considered in discrete time slots, this queue can be approximately characterized by a $DBMAP/D/1$ model for which the state transitions are described by the vector $\vec{x_i}$. Note that the load in this system is given by

$$\rho = \frac{n\bar{N}}{t} \tag{2.194}$$

where $\bar{N}$ is the average number of IP packets in a burst given by Eq. 2.185, and the arrival rate is given by $\lambda = 1/\bar{N}$.

Let P_i be the probability that i out of n sources change their state from ON to OFF in a time slot which can be approximately written under high load conditions and large number of sources as

$$P_i = \begin{cases} e^{-\lambda\rho} \approx 1 - \lambda\rho, & \text{for } i = 0 \\ \lambda\rho e^{-\lambda\rho} \approx \lambda\rho, & \text{for } i = 1 \\ \frac{(\lambda\rho)^i}{i!}e^{-\lambda\rho} \approx 0, & \text{otherwise,} \end{cases} \tag{2.195}$$

where the above approximation can be obtained with a first-order expansion of the binomial distribution for P_i. From Eq. 2.195, the probability that i IP packets arrive during the burst formation time, b_i, can be evaluated as

$$
b_i = \begin{cases} P_0 = 1 - \lambda\rho, & \text{for } i = 0 \\ P_i \pi_{i|(i \geq 1)} = \lambda^2 \rho (1 - \lambda)^{i-1}, & \text{for } i \geq 1. \end{cases} \tag{2.196}
$$

Note that b_i follows a geometric distribution. The transitions between the states of the burst queue are shown in Fig. 2.9. The transitions occur for each time slot in which b_i for any i indicates that i packets arrive during a time slot. The states indicate the number of packets in the burst transmission queue at any time.

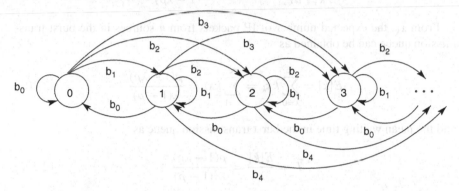

Fig. 2.9 Markov chain model for the burst transmission queue

From Fig. 2.9, a system of difference equations given below are derived for x_i as follows:

$$
x_0(1 - b_0) = x_1 b_0
$$
$$
x_1(1 - b_1) = x_2 b_0 + x_0 b_1
$$
$$
x_i(1 - b_1) = x_{i+1} b_0 + x_0 b_i + \sum_{k=1}^{i-1} x_i b_{i-k+1}, \quad \text{for } i \geq 2. \tag{2.197}
$$

The z-transform of the above system of equations gives

$$
(1 - b_1) \sum_{k=2}^{\infty} x_k z^k = \frac{b_0}{z} \sum_{k=2}^{\infty} x_{k+1} z^{k+1} + x_0 \sum_{k=2}^{\infty} b_k z^k + \sum_{k=2}^{\infty} \sum_{i=1}^{k-1} x_i b_{k-i+1} z^k \tag{2.198}
$$

from which the z-transform of x_i, $X(z)$ can be written as

$$X(z) = (1 - \rho) \left(1 - \frac{z\lambda\rho}{z(1 - \lambda) - (1 - \lambda\rho)} \right)$$
$$= x_0 \frac{zY(z) - Y(z)}{z - Y(z)}, \tag{2.199}$$

where $Y(z)$ is the z-transform of b_i given by

$$Y(z) = 1 - \lambda\rho + \lambda^2\rho \frac{z}{1 - (1 - \lambda)z}. \tag{2.200}$$

Since $X(1) = 1$, $x_0 = 1 - \rho$ which can be used to eliminate x_0 and obtain

$$x_k = \frac{1}{k!} \frac{d^k X(z)}{dz^k} \Big|_{z=0} = \frac{\lambda\rho(1 - \rho)(1 - \lambda)^{k-1}}{(1 - \lambda\rho)^k}. \tag{2.201}$$

From x_k, the expected number of IP packets from n sources in the burst transmission queue can be obtained as

$$E[k] = \sum_{k=0}^{\infty} kx_k = \sum_{k=1}^{\infty} kx_k = \frac{\rho(1 - \lambda\rho)}{\lambda(1 - \rho)} \tag{2.202}$$

and the mean waiting time in the burst transmission queue as

$$\bar{T} = \frac{E[k]}{\lambda} = \frac{\rho(1 - \lambda\rho)}{\lambda^2(1 - \rho)}. \tag{2.203}$$

Modeling the burst assembly queue along with the burst transmission queue helps to study the relationship between the burst formation time in the assembly queue and the waiting time at the transmission queue. The models for the burst assembly queue and the burst transmission queue considered in this section help to compute the end-to-end delay of the traffic considering the latency in the ingress node explicitly. The finite buffer model for the ingress node which gives the waiting time and the blocking probability of bursts also helps to dimension the transmission queue. Depending on the number of packets in a burst and the size of the burst assembly queue, an optimal size for the burst transmission queue can be derived which enables better dimensioning of the offset time. Such models are useful to determine the offset time accurately and send the control packet appropriately to reduce the end-to-end delay and improve the utilization of the bandwidth [28].

References

1. Baskett, F., Chandy, K.M., Muntz, R.R., Palacios, F.G.: Open, closed and mixed networks of queues with different classes of customers. Journal of the ACM 22(2), 248–260 (1975)
2. Baynat, B., Dallery, Y.: A product-form approximation method for general closed queueing network models of computer systems. Performance Evaluation 24(3), 165–188 (1996)

3. Chlamtac, I., Ganz, A.: Evaluation of the random token protocol for high-speed and radio networks. IEEE Journal on Selected Areas in Communications **5**(6), 969–976 (1987)
4. Fischer, W., Hellstern, K.M.: The Markov-modulated Poisson process (MMPP) cookbook. Performance Evaluation **18**(2), 149–171 (1993)
5. Ge, A., Callegati, F., Tamil, L.: On optical burst switching and self-similar traffic. IEEE Communications Letters **4**(3), 98–100 (2000)
6. Gross, D., Harris, C.: Fundamentals of Queueing Theory. Wiley-Interscience, USA (1997)
7. Hu, G.: Performance model for a lossless edge node of OBS network. In: Proceedings of IEEE GLOBECOM, pp. 1–6 (2006)
8. Hu, G., Dolzer, K., Gauger, C.: Does burst assembly really reduce self-similarity? In: Proceedings of OFC/NFOEC, pp. 124–126 (2003)
9. Izal, M., Aracil, J.: On the influence of self-similarity on optical burst switching traffic. In: Proceedings of IEEE GLOBECOM, pp. 2308–2312 (2002)
10. Kleinrock, L.: Queueing Systems. Vol. 1, John Wiley and Sons, USA (1975)
11. Laevens, K.: Traffic characteristics inside optical burst switched networks. In: Proceedings of SPIE Opticomm, pp. 137–148 (2002)
12. Lee, S.: An analytic model of burst queue at an edge optical burst switching node. In: Proceedings of Parallel and Distributed Computing: Applications and Technologies, pp. 459–463 (2004)
13. Lee, S.: Packet-based burst queue modeling at an edge in optical-burst switched networks. Computer Communications **29**(5), 634–641 (2006)
14. Marie, R.: An approximate analytical method for general queueing networks. IEEE Transactions on Software Engineering **5**(5), 530–538 (1979)
15. Marie, R.: Calculation of equilibrium properties for $\lambda(n)/C_k/1/N$ queues. ACM Sigmetrics Performance Evaluation Review **9**(2), 117–125 (1980)
16. Mountrouidou, X., Perros, H.G.: On the departure process of burst aggregation algorithms in optical burst switching. Computer Networks **53**(3), 247–264 (2009)
17. Neihardt, A.L., Wang, J.L.: The concept of relevant time scales and its application to queueing analysis of self-similar traffic. ACM Sigmetrics Performance Evaluation Review **26**(1), 222–232 (1998)
18. Neuse, D., Chandy, K.M.: HAM: The heuristic aggregation method for solving general closed queueing network models of computer systems. ACM Sigmetrics Performance Evaluation Review **11**(4), 195–212 (1982)
19. Perros, H.G.: Queueing Networks with Blocking: Exact and Approximate Solutions. Oxford University Press, USA (1994)
20. Racz, S., Jakabfy, T., Toth, G.: The cumulative idle time in an $ND/D/1$ queue. Queueing Systems: Theory and Applications **49**(3–4), 297–319 (2005)
21. Rodrigo, M.V., Gotz, J.: An analytical study of optical burst switching aggregation strategies. In: Proceedings of Workshop on Optical Burst Switching (2004)
22. Rostami, A., Wolisz, A.: Modeling and fast emulation of burst traffic in optical burst-switched networks. In: Proceedings of IEEE ICC, pp. 2776–2781 (2006)
23. Rostami, A., Wolisz, A.: Modeling and synthesis of traffic in optical burst-switched networks. Journal of Lightwave Technology **25**(10), 2942–2952 (2007)
24. Xu, L., Perros, H.G., Rouskas, G.N.: A queueing network model of an edge optical burst switching node. In: Proceedings of IEEE INFOCOM, pp. 2019–2029 (2003)
25. Yu, X., Chen, Y., Qiao, C.: Performance evaluation of optical burst switching with assembled burst traffic input. In: Proceedings of IEEE GLOBECOM, pp. 2318–2322 (2002)
26. Yu, X., Chen, Y., Qiao, C.: A study of traffic statistics of assembled burst traffic in optical burst switched networks. In: Proceedings of SPIE Opticomm, pp. 149–159 (2002)
27. Yu, X., Li, J., Cao, X., Chen, Y., Qiao, C.: Traffic statistics and performance evaluation in optical burst switched networks. Journal of Lightwave Technology **22**(12), 2722–2738 (2004)
28. Zalesky, A., Vu, H.L.: Designing an optimal scheduler buffer in OBS networks. Journal of Lightwave Technology **26**(14), 2046–2054 (2008)
29. Zeng, G., Chlamtac, I., Lu, K.: A finite queueing network model for burst assembler in OBS networks. In: Proceedings of SPECTS, pp. 642–648 (2004)

Chapter 3
Blocking Probability in OBS Networks

3.1 Introduction

Computing the blocking probability of bursts at the core nodes in an OBS network is important for many reasons. First, it helps to understand the loss process of bursts and provide exact analytical expression for the loss rate of bursts at the core nodes. Second, it helps to give bounds for the loss rate which is an important measure of QoS in OBS network. Finally, it helps to design new loss minimization mechanisms at various places in the network in order to improve the performance of the network at higher layers. Computation of blocking probability in OBS networks is challenging due to the lack of optical buffers at the core nodes and the unacknowledged nature of switching. Contention losses which are different from the traditional losses due to buffer overflow (congestion) lead to an additional complexity in formulating analytical expressions for blocking probability of bursts. In the circuit switching, arriving connection requests are blocked at the source due to lack of bandwidth. In packet switching, packets are buffered at the intermediate nodes if bandwidth is not available. Packets are dropped only when the buffer overflows. Unlike these two switching paradigms, which are analyzed with traditional queueing systems, losses in OBS network occur whenever two or more bursts request the same wavelength for the same duration. These losses are completely random and do not necessarily indicate congestion in the network. To a certain extent such contention is resolved either in the space domain by switching onto an alternate wavelength (wavelength conversion) or in the time domain by delaying the optical burst (using FDLs). When both these techniques cannot resolve the contention for a wavelength, bursts are dropped. The probability of bursts getting dropped due to contention at a node is commonly referred to as the blocking probability or the BLP.

In this chapter, analytical models for blocking probability of bursts are presented under different operational regimes. Initially, the discussion starts with simple models for blocking probability that treat loss at a core node as that of connection blocking in circuit-switched networks. Then models that are specific to OBS networks with JET reservation mechanism and with multiple traffic classes are presented which assume different settings. Blocking probability is evaluated when there are wavelength converters, with either limited or full conversion and also when there are

T. Venkatesh, C. Siva Ram Murthy, *An Analytical Approach to Optical Burst Switched Networks*, DOI 10.1007/978-1-4419-1510-8_3,
© Springer Science+Business Media, LLC 2010

no wavelength converters or a limited number of converters. The effect of offset time on the blocking probability is also discussed in detail. Finally, evaluation of blocking probability at a core node equipped with FDLs is discussed when the length of the FDLs is short and long.

3.2 Blocking Probability in JET/JIT-Based Reservation

Consider an OBS network that uses JET/JIT reservation protocol and the core nodes have full wavelength conversion capability but no optical buffers. In such a network, a burst is dropped at the core node when all the wavelengths are occupied for a part or the entire duration of the burst. Blocking at the core node in this system is modeled approximately using the theory of connection blocking in circuit switched networks [19]. Assume that the bursts arrive in an exponentially distributed way with a mean rate λ and the service rate of the bursts is also exponentially distributed with a mean $\mu = 1/L$ where L is the mean burst length. The traffic intensity (or load) is given by $\rho = \lambda/\mu$. In such a system the upper bound on the blocking probability (PB) of bursts at a node (outgoing link) with w wavelengths is approximated by the Erlang B formula of an $M/M/w/w$ queue [9, 38], i.e.,

$$PB(\rho, w) = E(\rho, w) = \frac{\frac{\rho^w}{w!}}{\sum_{i=0}^{w} \frac{\rho^i}{i!}}. \tag{3.1}$$

The load ρ is simply equal to λL for JET and $\lambda(L + \delta)$ for JIT since the wavelength is reserved only for the duration of the burst length L in JET while it is reserved even for the offset time δ in JIT. The increased load in JIT results in a higher loss probability compared to JET especially for larger values of offset time. The traffic load ρ is assumed to be uniform across all the wavelengths. The Erlang B formula for blocking is insensitive to the service time distribution which can be seen from its independence of μ. Thus the same formula holds for an $M/G/w/w$ queue also. If the core nodes are equipped with FDLs which can give a maximum delay of d units, the blocking probability can be approximated by modeling the system with an $M/M/w/d$ queue [14]. d is also the maximum number of bursts in a switch related to the number of FDLs f, by $d = w(1 + f)$. Blocking probability of such a system has a lower bound given by [38]

$$PB(\rho, w, d) = \frac{\rho^d}{w^{d-w}.w!} \cdot \left(\sum_{i=0}^{w-1} \frac{\rho^i}{i!} + \sum_{i=w}^{d} \frac{\rho^i}{w^{i-w}w!} \right)^{-1}. \tag{3.2}$$

Although, an $M/M/w/d$ queue is used to model the blocking probability of an OBS node with FDL, it is not exact because an FDL is fundamentally not same as a queue. The model is inaccurate for small values of d because the FDL has an upper limit on the amount of delay it can provide. For larger values of the delay, this model is fairly accurate. For larger burst lengths compared to the blocking time, this model predicts a lower loss. On the other hand for smaller burst lengths, it predicts a higher loss [38]. The next chapter analyzes the blocking probability at the nodes with FDLs using more accurate models.

3.3 Markov Chain Model for Blocking Probability

As mentioned in Section 3.2, the Erlang B loss formula of an $M/M/w/w$ queue is used to approximate the blocking probability of a core node in OBS networks with JET/JIT protocol. Though the expression is simple it is not accurate and is also not realistic. Particularly, it treats the output link as a loss system with offered load of λL which oversimplifies the system by neglecting the effect of offset time on the blocking probability. This discrepancy is larger with void filling algorithms because the Erlang B formula does not consider the effect of burst length (service duration) on the blocking probability. It considers only the average load which is the ratio of the mean arrival rate of bursts to the mean service rate. To overcome these shortcomings, this section discusses a Markov chain model that explicitly considers the effect of offset time and burst length [16–18]. The model is based on the classical theory of advanced reservation systems such as hotel reservation and tour planning. The JET protocol in OBS has similarity to the advanced reservation systems.

In this model, a wavelength is modeled using a discrete time Markov chain in which the time is divided into slots of width δ and it is assumed that offset time o and burst length l are integer multiples of δ. Let the BHP that arrives at time t_r request the wavelength for an interval $[t_r+o, t_r+o+l]$. If the interval requested does not overlap with an interval requested already, the burst is successfully transmitted otherwise the BHP and the corresponding burst are dropped. Burst segmentation and retro-blocking are ignored in this model. Figure 3.1 shows the slotted-time model used for JET protocol. t_a is the actual time when the data burst arrives which is equal to $t_r + o$. Assume that the arrival and departure of bursts always occur at the slot boundaries and the slots are numbered with respect to the arrival time slot of a reference BHP. Let the arrival of BHPs follow a Poisson distribution with mean rate of λ per time slot. The random variable O which represents the offset time has a probability distribution $f(o) = P(O = o)$ for $o = 0, 1, \ldots$, and similarly the distribution for data burst length (duration) $g(l) = P(L = l)$ for $l = 1, 2, \ldots$. For any time slot n, the number of requests from all previous slots including n is λ (from the definition of a Poisson process). Given a BHP arrival at slot 0, the probability that it requests a time slot n is given by $f(n)$ so that the total number of requests from 0th slot to the nth slot is $\lambda f(n)$. As seen by the BHP at slot 0 the number of data bursts arriving in slot n also consists of the number of requests from all time slots before 0 in addition to those from slot 0. Thus the number of arrivals in slot n as seen by a reference BHP is the sum of all possible requests which is λ minus any requests to be made in future for slot n from slots 1 to n. Therefore, the number of

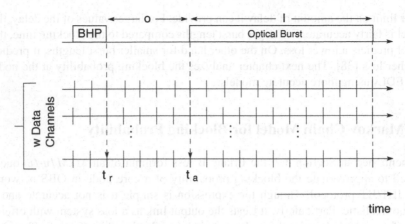

Fig. 3.1 Illustration of slotted-time model for JET

data bursts whose start time is within slot n is Poisson distributed with mean

$$\lambda_{(n)} = \begin{cases} \lambda, & \text{for } n \le 0 \\ \lambda\{1 - \sum_{o=0}^{n-1} f(o)\}, & \text{for } n > 0. \end{cases} \tag{3.3}$$

Assume that the duration of the data burst has a geometric distribution, i.e., $g(l) = q(1-q)^l$ with mean $\overline{L}$ and $q = 1/\overline{L}$. Let $S^{(n)}$ be the state of a wavelength in time slot n such that $S^{(n)} = 1$ if the wavelength is occupied, otherwise $S^{(n)} = 0$. Assuming that bursts arrive uniformly across all the w wavelengths so that the arrival rate on a single wavelength is given by $\hat{\lambda} = \lambda/w$. With these assumptions, a single wavelength can be modeled as a non-homogeneous Markov chain with a transition probability matrix $M^{(n)}$ given by

$$M^{(n)} = \begin{bmatrix} p_{0,0}^{(n)} & p_{0,1}^{(n)} \\ p_{1,0}^{(n)} & p_{1,1}^{(n)} \end{bmatrix}, \tag{3.4}$$

where the transition probabilities are defined as

$$p_{i,j}^{(n)} = P(S^{(n+1)} = j \mid S^{(n)} = i), \quad i, j = 0, 1$$

$$= \begin{cases} e^{-\hat{\lambda}_{(n+1)}}, & i = j = 0 \\ 1 - e^{-\hat{\lambda}_{(n+1)}}, & i = 0, j = 1 \\ qe^{-\hat{\lambda}_{(n+1)}}, & i = 1, j = 0 \\ 1 - qe^{-\hat{\lambda}_{(n+1)}}, & i = j = 1. \end{cases} \tag{3.5}$$

Next, let the number of wavelengths reserved on an output link in a slot n be denoted by $Z^{(n)} \in [0, 1, \ldots, w]$. The state of the output link, which is defined in terms of the number of busy wavelengths, can also be modeled as a non-homogeneous Markov chain with transition probability matrix $R^{(n)}$ given by

$$R^{(n)} = \begin{bmatrix} r_{0,0}^{(n)} & r_{0,1}^{(n)} & \cdots & r_{0,w}^{(n)} \\ r_{1,0}^{(n)} & r_{1,1}^{(n)} & \cdots & r_{1,w}^{(n)} \\ \vdots & \vdots & \ddots & \vdots \\ r_{w,0}^{(n)} & r_{w,1}^{(n)} & \cdots & r_{w,w}^{(n)} \end{bmatrix} \tag{3.6}$$

with the transition probabilities defined as

$$
\begin{aligned}
r_{i,j}^{(n)} &= P(Z^{(n+1)} = j \mid Z^{(n)} = i), \quad i,j = 0, 1, \ldots, w \\
&= \begin{cases} \sum_{z=0}^{i} \binom{i}{z} q^z (1-q)^{i-z} e^{-\lambda_{(n+1)}} \frac{\lambda_{(n+1)}^{j-i+z}}{(j-i+z)!}, & j \le w-1, i \le j \\ \sum_{z=0}^{j} \binom{i}{(i-j+z)} q^{i-j+z} (1-q)^{j-z} e^{-\lambda_{(n+1)}} \frac{\lambda_{(n+1)}^{z}}{z!}, & j \le w-1, i \ge j \\ 1 - \sum_{z=0}^{w-1} r_{i,z}^{(n)}, & j = w. \end{cases}
\end{aligned}
\tag{3.7}
$$

The stationary probabilities for $Z^{(n)}$ are given by

$$\Pi = \begin{bmatrix} \pi_0 & \pi_1 & \ldots & \pi_w \end{bmatrix} \tag{3.8}$$

which can be obtained by solving the set of equations below

$$\pi_j = \sum_{i=0}^{w} r_{i,j}^{(0)} \cdot \pi_i, \quad 0 \le j \le w. \tag{3.9}$$

To calculate the blocking probability of a data burst with duration l, the state of the wavelength for all slots $n = o, o+1, \ldots, o+l-1$ needs to be considered. If the probability distribution of $Z^{(n)}$ is denoted by $v_z^{(n)}$, $z = 0, 1, \ldots, w$, then a unit length data burst is blocked with a probability $v_w^{(o)}$ where o is the offset time. Let $T_s(o, l)$ be the probability that the system remains in state $S^{(n)}$ for a duration l starting from slot o. The probability that a request for reservation with the parameters $(o.l)$ is accepted is equal to the probability that at least one wavelength is free for the interval $[o, o + l]$. Correspondingly the blocking probability for such a data burst can be written as

$$PB(o, l) = 1 - T_0(o, l) \sum_{z=0}^{w-1} v_z^{(o)}, \tag{3.10}$$

where the probability $v_z^{(o)}$ is given by

$$v_z^{(o)} = \Pi R^{(0)} R^{(1)} \ldots R^{(o-2)} R_{(:,z)}^{(o-1)}. \tag{3.11}$$

The last term $R_{(:,z)}^{(o-1)}$ denotes the zth column of the matrix $R^{(o-1)}$. In Eq. 3.10, $T_0(o, l)$ indicates the probability that the wavelength is free for the duration of the burst which can be expressed as

$$T_0(o, l) = p_{0,0}^{(o)} p_{0,0}^{(o+1)} \cdots p_{0,0}^{(o+l-2)}$$
$$= e^{-\hat{\lambda}_{o+1}} e^{-\hat{\lambda}_{o+2}} \cdots e^{-\hat{\lambda}_{o+l-1}}. \tag{3.12}$$

To simplify the evaluation of the above equation, the offset time may be assumed to have a uniform distribution in the interval $[0, t_{max}]$ so that Eq. 3.3 can be written as

$$\lambda_{(n)} = \begin{cases} \lambda, & \text{for } n \le 0 \\ \lambda\{1 - \frac{n}{1+t_{max}}\}, & \text{for } 0 < n \le t_{max} \\ 0, & \text{for } n > t_{max}. \end{cases} \tag{3.13}$$

Imposing the condition that $o + l \le t_{max} + 1$, Eq. 3.12 can be written as

$$T_0(o, l) = e^{-\hat{\lambda}\{1 - \sum_{i=0}^{o} \frac{1}{1+t_{max}}\}} e^{-\hat{\lambda}\{1 - \sum_{i=0}^{o+1} \frac{1}{1+t_{max}}\}} \cdots e^{-\hat{\lambda}\{1 - \sum_{i=0}^{o+l-2} \frac{1}{1+t_{max}}\}}$$
$$= e^{-\hat{\lambda}[(1-k(o+1))+(1-k(o+2))+\cdots+(1-k(o+l-2))]}$$
$$= e^{-\hat{\lambda}\sum_{j=1}^{l-1}[1-k(o+j)]}$$
$$= e^{-\hat{\lambda}[(1-ko)(l-1)-k(l-1)l/2]}$$
$$= e^{-\hat{\lambda}(l-1)[1-k(o+l/2)]}, \tag{3.14}$$

where $k = \frac{1}{1+t_{max}}$. In general Eq. 3.10 gives the blocking probability of bursts in a JET-based OBS system including the effect of offset time and burst duration. As shown here for the case of uniformly distributed offset times, blocking probability can be obtained for any general distribution of offset times and burst length. However, the analysis uses a slotted-time model which assumes that the offset time is an integer multiple of a basic unit of time slot. The slotted-time model assumes that processing time is fixed and that the burst length and offset time can be expressed as multiples of the processing time. This prevents analysis of the effect of continuous values for offset time. In the next section, this assumption is relaxed and the effect of variable offset time on the blocking probability is discussed.

3.4 Blocking Probability with Variable Offset Time

In this section the effect of varying the offset time on the blocking probability of the data bursts is analyzed. In some models for blocking probability the effect of offset time is included either directly or indirectly through the definition of load at the core nodes. However, most of them assume that the offset times are equal for all the bursts. Some proposals which use different offset times for different classes of traffic also assume the offset time to be fixed and some integer multiple of the hop length. To a major extent the offset time is determined by the number of hops along the path multiplied by a constant factor representing the processing time for the BHP. But since the BHPs may be queued along the path and the number of voids to be searched for a feasible schedule is variable, the processing time at each node and the waiting time of the BHP depend on the load at the node. Therefore, assume that the offset times are constant or integer multiples of a fixed processing time do not give accurate blocking probability. Further, it does not capture the effect of variable burst length and different burst length distributions on the blocking probability. This section evaluates the blocking probability of bursts at a core node on a single wavelength as a function of continuous-time variable offset values [15]. However, no additional offset time used to provide QoS is assumed in this analysis. The offset time of bursts is computed only based on the number of hops to the destination node.

To compute the blocking probability of data bursts, it is enough to compute the number of bursts dropped due to the lack of a feasible schedule. A data burst cannot be scheduled if the duration of the burst is larger than the void created by a previous schedule or there is no free wavelength at the time of its arrival (assuming that it cannot be delayed). Consider the schedule on a single wavelength as shown in Fig. 3.2. Let the time of arrival of the BHP under consideration (indicated as current BHP in the figure) be denoted by $t = 0$ and the offset time be denoted by the interval $[\delta, \delta + O]$, where δ is the BHP processing time. The time interval $[\delta, \delta + O]$ is also known as the *horizon time*. As shown in Fig. 3.2 the interval $[-O, 0]$ is called as past-horizon time which is essentially of the same duration as the horizon time. The past-horizon time is important because the successful reservations made for the duration of horizon time originate at that time. To compute the probability of a burst not finding a suitable schedule in the horizon time, the number of BHPs arrived in the past-horizon time and the number of attempts for scheduling during the horizon time must be computed [15].

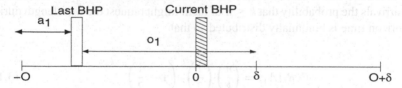

Fig. 3.2 Horizon time illustrated on a single wavelength

Assume that the BHPs and the data bursts arrive following a Poisson process with a mean arrival rate λ. Assume that the data bursts are of constant length typically generated by a size-based assembly mechanism at the edge node. Let $a_i, i = 1, 2, \ldots, n$ denote the arrival time of the ith BHP and o_i denote the corresponding offset time. Since the data bursts arrive with a Poisson distribution, given that there are n BHP arrivals in the past-horizon time $[-O, 0]$, both a_i and o_i are uniformly distributed in the intervals $[-O, 0]$ and $[\delta, \delta + O]$, respectively. The next step in evaluating the probability of a given BHP being blocked is to evaluate the

number of BHPs requesting the data bursts to be scheduled in the horizon time of
the current BHP and the number of them successfully scheduled.

To derive the probability distribution of the number of BHPs that requested the
corresponding data bursts to be scheduled in the horizon time of a given BHP, note
that only those BHPs that arrive during $[-O, 0]$ may attempt to reserve the wavelength during $[\delta, \delta + O]$. Since it is assumed that the offset times are uniformly
distributed in the interval $[\delta, \delta + O]$, only the BHPs during the past-horizon time
need to be considered. Let a hypothetical BHP arrive at time $a_1 = -O$ with an offset
time of O and a processing time of δ so that the data burst needs to be scheduled
after time $-O+(O+\delta) = \delta$. Similarly, all the BHPs that arrive up to the time before
that under consideration, i.e., before time $t = 0$, would attempt reservation in the
horizon time $[\delta, \delta + O]$. Therefore, to consider the blocking of a BHP that arrives
at $t = 0$, it is enough to consider the number of requests for reservation possible
during $[-O, 0]$. Note that it is also assumed that the burst sizes are not larger than
the offset times, otherwise those requests before $-O$ might also block the requests
in the horizon time. Assuming that there is only one arrival in the past-horizon time,
indicated by say A_1, the probability that it requests the wavelength during the hori-
zon time of the current BHP, $P(R_1|A_1)$, is simply the probability that $a_1 + o_1 > \delta$.
Therefore,

$$
\begin{aligned}
P(R_1|A_1) &= \int_{-O}^{0} Pr(o_1 > \delta - a_1) f_{a_1}(y) dy \\
&= \int_{-O}^{0} \frac{-y}{O} \frac{1}{O} dy = \frac{1}{2},
\end{aligned}
\tag{3.15}
$$

where $f_{a_1}(y)$ is the pdf of the arrival time. This equation shows that only half of the
arrivals in the past-horizon time request the wavelength during the current horizon
time. Since the BHPs are assumed to arrive in a uniform and independent manner,
for n arrivals the probability that $k \leq n$ of them might request the wavelength during
the horizon time is binomially distributed so that

$$
P(R_k|A_n) = \binom{n}{k} \left(\frac{1}{2}\right)^k \left(1 - \frac{1}{2}\right)^{n-k}.
\tag{3.16}
$$

Therefore, the number of requests for the wavelength during $[\delta, \delta + O]$ assuming
that there are n arrivals with mean rate λ is given by

$$
\begin{aligned}
P(R_k) &= \sum_{n=k}^{\infty} \binom{n}{k} \left(\frac{1}{2}\right)^k \left(1 - \frac{1}{2}\right)^{n-k} \frac{(\lambda O)^n}{n!} e^{-\lambda O} \\
&= \frac{(\frac{\lambda O}{2})^k}{k!} e^{-\lambda O} \sum_{n=k}^{\infty} \frac{(\frac{\lambda O}{2})^{n-k}}{(n-m)!} = \frac{(\frac{\lambda O}{2})^k}{k!} e^{-\lambda/2O},
\end{aligned}
\tag{3.17}
$$

which shows that the number of requests for the wavelength follows Poisson distribution with a mean $\lambda/2$.

To extend the same analysis for any interval other than $[\delta, \delta + O]$ considered above, let there be m_p portions of size $O_p = O/m_p$ in the interval $[\delta, \delta + O]$. Note that, for an interval of $[\delta + mO_p, \delta + (m+1)O_p], m = 0, 1, \ldots, m_p - 1$ during the horizon time of the current BHP, only those BHPs that arrive during $[mO_p - O, 0]$ of the past-horizon time might request the wavelength. Such an event occurs with a probability

$$
\begin{aligned}
r_m &= \int_{(m+1)O_p-O}^{0} Pr(\delta + mO_p - a_1 \le o_1 \le \delta + (m+1)O_p - a_1) f_{a_1}(y) dy \\
&\quad + \int_{mO_p-O}^{(m+1)O_p-O} Pr(o_1 \ge mO_p - a_1) f_{a_1}(y) dy \\
&= \int_{(m+1)O_p-O}^{0} \frac{O_p}{O} \frac{1}{O} dy + \int_{mO_p-O}^{(m+1)O_p-O} \frac{1}{O} \frac{O-(mO_p-y)}{O} dy \\
&= \frac{O_p}{O} - \frac{2m+1}{2} \left(\frac{O_p}{O}\right)^2.
\end{aligned} \tag{3.18}
$$

Note that $\sum_{m=0}^{m_p-1} r_m = 1/2$. Using the formula for r_m it can be shown that the number of requests for reservation during $[\delta + mO_p, \delta + (m+1)O_p]$ is given by

$$
P(R_k, [\delta + mO_p, \delta + (m+1)O_p]) = \frac{(\lambda r_m O)^k}{k!} e^{-\lambda r_m O}, \tag{3.19}
$$

which shows that it follows a Poisson distribution with rate λr_m.

After computing the number of requests for reservation during the horizon time, the probability of finding a successful reservation is computed to evaluate the blocking probability of the current BHP. For a BHP with horizon time $[\delta, \delta + O]$, the probability of a request finding the wavelength free for the entire duration depends on the number of attempts for reservation made. Let r be the number of attempts made for reservation of the wavelength during the horizon time and let $(V_1, V_2, \ldots, V_r)$ be the r-dimensional random variable indicating the sorted arrival times of all such requests. The joint-probability distribution of $(V_1, V_2, \ldots, V_r)$ can be obtained from the order statistics of r uniformly distributed arrivals over a period of O as [7]

$$
f_{V_1, V_2, \ldots, V_r}(z_1, z_2, \ldots, z_r) = \frac{r!}{O^r}, \tag{3.20}
$$

where $z_i \in [0, O], i = 1, 2, \ldots, r$ indicates the actual burst arrival time $z_i + \delta$. Let M be the service time of the burst and $P(S_r)$ be the probability of r successful attempts of reservation made during the horizon time $[\delta, \delta + O]$. $P(S_r)$ gives the probability that none of the attempts for reservation of the wavelength described by the process R_k overlap with another one. For $r = 1$, $P(S_r)$ can be obtained from Eq. 3.20 as

$$P(S_1) = \int_0^O \frac{1}{O} dz_1 = 1, \tag{3.21}$$

which is intuitive since the probability of a single arrival being successful is always unity. When there are two requests for reservation, i.e., $r = 2$ the two bursts overlap if $z_2 < z_1 + M$ so that the probability of two requests not overlapping is

$$P(S_2) = \int_0^{z_2-M} dz_1 \int_M^O \frac{2!}{O^2} dz_2$$
$$= \frac{2!}{O^2} \int_M^O (z_2 - M) dz_2 = \left(\frac{O-M}{O}\right)^2. \tag{3.22}$$

For $r = 3$,

$$P(S_3) = \int_0^{z_2-M} dz_1 \int_M^{z_3-M} dz_2 \int_{2M}^O \frac{3!}{O^3} dz_3$$
$$= \frac{3!}{O^3} \int_M^{z_3-M} (z_2 - M) dz_2 \int_{2M}^O dz_3 = \left(\frac{O-2M}{O}\right)^3. \tag{3.23}$$

For any r scheduling attempts, the non-overlapping probability among them is given by

$$P(S_r) = \left(\frac{O - (r-1)M}{O}\right)^r. \tag{3.24}$$

Therefore, assuming that r data bursts have been successfully scheduled in the horizon time of the current BHP, the probability of scheduling the current BHP successfully is given by

$$P(S_{r+1}|S_r) = \frac{P(S_{r+1})}{P(S_r)} = \frac{\left(\frac{O-rM}{O}\right)^{r+1}}{\left(\frac{O-(r-1)M}{O}\right)^r}$$
$$= \frac{1}{O} \frac{(O-rM)^{r+1}}{(O-(r-1)M)^r}, r = 0, 1, \ldots. \tag{3.25}$$

Note that the above equation uses $P(S_{r+1} \cap S_r) = P(S_{r+1})$ since S_{r+1} automatically implies S_r and the size of the burst is smaller than the offset time as implied by the condition $rM \leq O$.

Combining all the equations developed till now for the number of attempts made for scheduling and the number of those that are successful among them gives the blocking probability of a given BHP. Starting with a Poisson distribution for the arrival of BHPs in an interval of time, the number of BHPs (say k) that requests the wavelength over the horizon time of the current BHP is given by Eq. 3.17. Among these arrivals, some of them (say $k - r$) would be blocked with a probability say B and the remaining ones are successfully scheduled. The probability that r requests are successfully scheduled is given by Eq. 3.24 and the probability of the next request being successful is given by Eq. 3.25. Therefore, the blocking probability of a BHP can be written from all of these as

$$B = 1 - \sum_{k=0}^{\infty} \sum_{r=0}^{k} \binom{k}{r} (1 - B)^r B^{k-r} P(S_{r+1}|S_r) \frac{(\frac{\lambda}{2}O)^k}{k!} e^{-\lambda/2O}. \qquad (3.26)$$

The equation on the right-hand side evaluates the probability of a burst being blocked when r bursts have already been scheduled on the wavelength during the horizon time of the current BHP out of k requests. Note that the blocking probability of the current BHP is evaluated considering that the $k - r$ bursts out of k arrivals are blocked with the same probability B. This is possible only when all the bursts have offset times in the range $[\delta, \delta + O]$, i.e., with no restriction in the offset time values. The assumption of uniformly distributed offset times is validated by simulations in [15].

As considered earlier, for any general range of offset values, i.e., for any interval $[\delta + mO_p, \delta + (m + 1)O_p] \in [\delta, \delta + O]$, $m = 0, 1, \ldots, m_p - 1$ of the horizon time, the distribution of the number of attempts for reservation given by Eq. 3.19 can be used. Assuming that there are k arrivals over the interval $[\delta + mO_p, \delta + (m + 1)O_p]$ out of which $k - r$ have been blocked with a probability B, the probability of the current BHP being blocked can be approximated by

$$B_m \approx 1 - \sum_{k=0}^{\infty} \sum_{r=0}^{k} \binom{k}{r} (1 - B)^r B^{k-r} P_m(S_{r+1}|S_r) \cdot \frac{(\lambda r_m O)^k}{k!} e^{-\lambda r_m O}, \qquad (3.27)$$

where r_m is defined by Eq. 3.18 and

$$P_m(S_{r+1}|S_r) = \frac{1}{O_p} \frac{(O_p - rM)^{r+1}}{(O_p - (r - 1)M)^r}. \qquad (3.28)$$

Assuming that the burst size is small enough so that none of the bursts scheduled in the past-horizon time block the bursts scheduled in the current horizon time helps in making the analysis tractable. This assumption is validated by simulations in [15]. The analysis is quite good for low load (therefore low blocking probabilities) and helps in equalizing the blocking probabilities of bursts with different offset values. Any method of proactively discarding the bursts to prevent unfairness among bursts with different offset times can be analyzed using this approach.

3.5 Blocking Probability with and Without Wavelength Conversion

As pointed out earlier, wavelength conversion is one of the methods used to resolve contention in OBS networks. In full wavelength conversion, a burst that arrives on a particular wavelength is switched onto another wavelength on the output link if it is available for the burst duration. It is demonstrated that, compared to the case without wavelength conversion, where the burst must be switched on the same wavelength, full wavelength conversion reduces the BLP significantly [5, 21]. In this section a simple analytical model that uses a finite state model for an OBS core node both with and without wavelength conversion capability is discussed [31]. The model helps in the evaluation of blocking probability and the steady-state throughput at a core node which is the average number of bursts received successfully over a unit time. Since the technology for wavelength conversion is still immature and expensive, a model that can evaluate the blocking probability with and without wavelength conversion can be used to evaluate the benefit of increasing the number of converters compared to increasing the number of wavelengths. In simple terms, the model helps in answering if increasing the number of wavelengths is better than using wavelength converters.

Consider the transmission of a burst in a JET-based OBS network as shown in Fig. 3.3(a) and the corresponding representation of the slotted-time model where the entire duration of the burst transmission (t), starting from the arrival of the BHP till the transmission of the data burst, is divided into small time slots. Let t_i be the duration of each slot in total time t, and let the total number of slots $l = t/t_i$ be an integer. Assume that bursts arrive following a Poisson distribution with mean λ and the t_i is small enough that only one burst arrives in a slot, i.e., the probability that two or more bursts arrive in a slot of duration t_i is zero. All bursts are assumed to be of fixed length and it takes l slots to transmit an accepted burst. However, the analysis can be used for bursts of variable length considering l to be the average service time of bursts. With these assumptions, the following sections provide a model for a core node for the cases of no-wavelength conversion and full wavelength conversion.

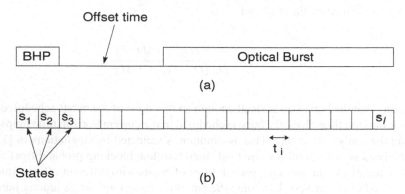

Fig. 3.3 Transmission of an optical burst and the corresponding slotted-time model

3.5.1 Model for Node Without Wavelength Conversion

To facilitate easy understanding of the model, consider a simple case where the burst length occupies only three slots or $l = 3$ and the number of wavelengths $w \geq l$. The state diagram that models the burst transmission in this case is given in Fig. 3.4. The state space consists of four sets of states with the initial state represented by s_0 indicating that the node is idle (no burst being serviced). The node remains in this state as long as there is no new arrival of burst which occurs with a probability $1 - p_a$, where p_a is the probability that a new burst arrives. When a new burst arrives, the state changes to s_1 which is a part of the next set of states denoted by the set $S_1 = \{s_1, s_2, \ldots, s_l\}$.

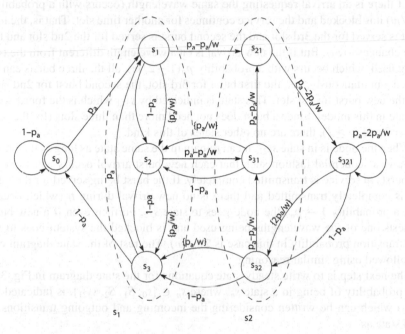

Fig. 3.4 State diagram for an OBS node with $l = 3$ and $w \geq l$. The probabilities shown within braces denote the blocking probabilities

In the second set of states (i.e., a state in the set S_1), only one of the wavelengths is being used and there is a transition among the states in S_1 for each time slot. For example, when the node is in state s_1 and there is no new arrival during the interval t_i or if the new burst gets blocked because it requests the same wavelength, the system moves to state s_2. The first event occurs with a probability $1 - p_a$ while the second one occurs with a probability p_a/w (blocking probability). When the system remains in the state s_2 for another slot either without a new arrival or a new arrival on the same wavelength gets blocked, it goes to state s_3. In the state s_3, if there is no new arrival for t_i the node moves to the idle state indicating the completion of the transmission. But if a new burst arrives during t_i (occurs with probability p_a), it is served on the same wavelength indicated by the state s_1. Note that only three equivalent states are shown in the set S_1 because $l = 3$. When the node is in either s_1 or s_2 and there is a new arrival on another wavelength (occurs with a probability $p_a(1 - 1/w)$), the system goes to s_{21} and s_{31}, respectively.

These states which are part of the third set of states denoted by $S_2 = \{s_{ij}\}$, i, $j \in \{1, 2, 3\}$ and $i > j$ indicate that the node is serving ith slot of the first burst and jth slot of the second burst. For example, the system goes to s_{21} when a new burst arrives on a wavelength different from the one serving the first burst. So while the second slot of the first burst is being served, the first slot of the second burst can be served by the other wavelength. In this case, since a burst can be served in three time slots, there is no transition from s_3 to a state in the set S_2. When the node is in state s_{21}, if there is an arrival requesting the same wavelength (occurs with a probability $2p_a/w$) it is blocked and the service continues for another time slot. That is, the first burst is served for the 3rd slot and the second burst is served for the 2nd slot and the state changes to s_{32}. But if the new arrival is on a wavelength different from the two being used, which occurs with a probability $p_a(1 - 2/w)$, all the three bursts can be served simultaneously, i.e., the first burst for 3rd slot, the second burst for 2nd slot, and the new burst for 1st slot. This state is indicated by s_{321} which is the fourth kind of state in this model. Since a burst does not need more than three slots (for the case of $l = 3$ and $w \geq l$), there are no other states of this kind.

When the node is in state s_{321}, it can remain in the same state as long as the bursts arrive in a sequential fashion such that each new burst arrival occurs when one of the bursts in service is transmitted completely. If the burst being served for the third slot is completely transmitted and there is no new arrival during t_i (which occurs with a probability $1 - p_a$) the node goes to state s_{32}. Further, even if a new burst requests one of the wavelengths being used and is blocked the system goes to s_{32} (the transition probability in this case is $2p_a/w$). The rest of the state diagram can be followed using similar reasoning.

The next step is to write steady-state equations for the state diagram in Fig. 3.4. The probability of being in a state s_n, where $s_n \in \{s_0, S_1, S_2, s_{321}\}$, is indicated by $P(s_n)$ which can be written considering the incoming and outgoing transitions at each state as

$$P(s_1) = p_a(P(s_0) + P(s_3))$$

$$P(s_2) = \left(1 - p_a + \frac{p_a}{w}\right)(P(s_1) + P(s_{31}))$$

$$P(s_3) = \left(1 - p_a + \frac{p_a}{w}\right)(P(s_2) + P(s_{32}))$$

$$P(s_{21}) = \left(p_a - \frac{p_a}{w}\right)(P(s_1) + P(s_{31}))$$

$$P(s_{31}) = \left(p_a - \frac{p_a}{w}\right)(P(s_2) + P(s_{32}))$$

$$P(s_{32}) = \left(1 - p_a + \frac{2p_a}{w}\right)(P(s_{21}) + P(s_{321}))$$

$$P(s_{321}) = \left(p_a - \frac{2p_a}{w}\right)(P(s_{321}) + P(s_{21})). \tag{3.29}$$

Solving the above set of equations gives the probability of being in a state as

$$P(s_1) = P(s_2) = P(s_3) = \frac{p_a}{1 - p_a} \cdot P(s_0) = \frac{p_a w}{w(1 - p_a)} \cdot P(s_0)$$

$$P(s_{21}) = P(s_{31}) = P(s_{32}) = \frac{(w - 1)p_a^2}{(w - (w - 1)p_a)(1 - p_a)} \cdot P(s_0)$$

$$= \frac{w}{w\frac{(1 - p_a)}{p_a}} \cdot \frac{w - 1}{w\frac{(1 - p_a)}{p_a} + 1} \cdot P(s_0)$$

$$P(s_{321}) = \frac{(w - 1)(w - 2)p_a^3}{(w - (w - 2)p_a)(w - (w - 1)p_a)(1 - p_a)} \cdot P(s_0)$$

$$= \frac{w}{w\frac{(1 - p_a)}{p_a}} \cdot \frac{w - 1}{w\frac{(1 - p_a)}{p_a} + 1} \cdot \frac{w - 2}{w\frac{(1 - p_a)}{p_a} + 2} \cdot P(s_0).$$

$$(3.30)$$

The probability of being in the initial state can be obtained solving the set of equations in Eq. 3.30 along with the fact that the sum of all probabilities equals one, i.e.,

$$P(s_0) + 3P(s_1) + 3P(s_{21}) + P(s_{321}) = 1. \tag{3.31}$$

Therefore, $P(s_0)$ is given by

$$P(s_0) =$$
$$\left[1 + 3 \cdot \frac{w}{w\frac{(1 - p_a)}{p_a}} + 3 \cdot \frac{w}{w\frac{(1 - p_a)}{p_a}} \cdot \frac{w - 1}{w\frac{(1 - p_a)}{p_a} + 1} + \frac{w}{w\frac{(1 - p_a)}{p_a}} \cdot \frac{w - 1}{w\frac{(1 - p_a)}{p_a} + 1} \cdot \frac{w - 2}{w\frac{(1 - p_a)}{p_a} + 2} \right]^{-1}.$$

Note that in the above model, the number of wavelengths is large enough to accommodate the transmission of bursts for three time slots. In general, if $w < l$ the same model can be used after some simple modifications. If $w = 2$, the state s_{321} does not exist so that $P(s_{321}) = 0$. Similarly if $w = 1$, $P(s_{21}) = P(s_{31}) = P(s_{32}) = P(s_{321})) = 0$. The probability of the initial state can also be evaluated accordingly. In a similar way the model can be extended for larger values of l and for either the case of $w > l$ or $w < l$.

The advantage of this model is that it can be used to compute the steady-state blocking probability which is the probability that a new burst is blocked, and the steady-state throughput which is defined as the average number of bursts received successfully per unit time. The steady-state throughput can be defined in general as

$$\tau(p_a, w, l) = \sum_{i=1}^{l} P(s_i) + \sum_{i=j+1}^{l} \sum_{j=1}^{l-1} 2P(s_{ij}) + \sum_{i=j+1}^{l} \sum_{j=k+1}^{l-1} \sum_{k=1}^{l-2} 3P(s_{ijk}) + \cdots$$

$$+ \sum_{i_l=i_{l-1}+1}^{l} \sum_{i_l=i_{l-2}+1}^{l-1} \cdots \sum_{i_2=1}^{2} (l-1)P(s_{i_l i_{l-1} \ldots i_2}) + l P(s_{l(l-1) \ldots 1}). \tag{3.32}$$

The definition will be clearer when the generalization of the above model is done for any value of l and w but, temporarily it can be used to evaluate the steady-state throughput for the case of $l = 3$ as

$$\tau(p_a, w, 3) = \begin{cases} 3(P(s_1) + 2P(s_{21}) + P(s_{321})), & \text{if } w \geq 3 \\ 3(P(s_1) + 2P(s_{21})), & \text{if } w = 2 \\ 3P(s_1), & \text{if } w = 1. \end{cases} \tag{3.33}$$

Similarly, the steady-state blocking probability can be evaluated by adding all the blocking probabilities indicated in Fig. 3.4 (by probabilities marked in braces) after multiplying with the corresponding steady-state probabilities. Therefore, the blocking probability for the case of $l = 3$ can be written as

$$\text{PB} = P(s_1).\frac{p_a}{w} + P(s_2).\frac{p_a}{w} + P(s_{21}).\frac{2p_a}{w} + P(s_{31}).\frac{p_a}{w} + P(s_{32}).\frac{p_a}{w} + P(s_{321}).\frac{2p_a}{w}$$

$$= \frac{p_a}{w}.2(P(s_1) + 2P(s_{21}) + P(s_{321})) = \frac{2p_a}{3w}\tau(p_a, w, 3). \tag{3.34}$$

3.5.1.1 General Model for No-Wavelength Conversion

Consider an OBS network without wavelength conversion capability at the core nodes with w wavelengths in each link and bursts that have fixed length which can be served in $l \geq 1$ time slots each of duration t_i. For this general case, the state diagram discussed in Fig. 3.4 can be extended with the nth state denoted by $S_n = s_{i_n, i_{n-1}, \ldots, i_1}$ where $n \in \{1, 2, \ldots, \min\{l, w\}\}$ and $i_1 < i_2 <, \ldots, < i_n \in \{1, 2, \ldots, l\}$. As mentioned earlier for the simpler case, the state $s_{i_n, i_{n-1}, \ldots, i_1}$ indicates that at the core node, the first burst is being served for the ith$_n$ slot, the second burst for the ith$_{n-1}$ slot, and so on. Depending on the status of the node at the time of a new arrival, there can be two cases to arrive at the state S_n which are discussed below. These two cases of transition to S_n are shown in Figs. 3.5 and 3.6 with the corresponding transition probabilities.

- $i_1 = 1$: A new burst arrives and finds that there is a free wavelength so that it is served in the first slot. Under this condition, the previous states can be either $S_{n-1} = s_{i_{n-1}, i_{n-1}-1, \ldots, i_2-1}$ or $S_n = s_{l, i_n-1, i_{n-1}-1, \ldots, i_2-1}$ (see Fig. 3.5). The transition probability from these states is the probability of a new arrival combined with the probability of the burst selecting an unused wavelength which can be written as

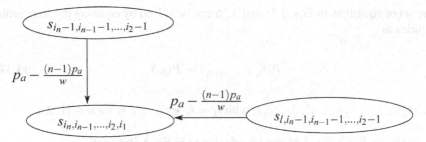

Fig. 3.5 Generic representation of nth state for an OBS node without wavelength conversion: $i_1 = 1$ and $(w \geq l \geq 1$ or $n \neq w)$

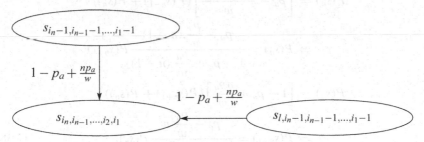

Fig. 3.6 Generic representation of nth state for an OBS node without wavelength conversion: $i_1 \neq 1$ and $(w \geq l$ or $n \neq w)$

$$p_a \left[1 - \frac{(n-1)}{w} \right] = p_a - \frac{(n-1)p_a}{w}.$$

The set of steady-state equations can be written as

$$P(s_{i_n,i_{n-1},\ldots,i_2,i_1}) = \left[p_a - \frac{(n-1)p_a}{w} \right]$$
$$\times \left(P(s_{i_n-1,i_{n-1}-1,\ldots,i_2-1}) + P(s_{l,i_n-1,i_{n-1}-1,\ldots,i_2-1}) \right). \quad (3.35)$$

- $i_1 \neq 1$: Here, either there is no arrival of a burst or a new burst arrives to find the requested wavelength to be busy and is blocked. The previous states in this case can be either $S_n = s_{i_n-1,i_{n-1}-1,\ldots,i_1-1}$ or $S_{n+1} = s_{l,i_n-1,i_{n-1}-1,\ldots,i_1-1}$ (see Fig. 3.6). The transition probability from these states is the sum of probability of no arrivals and the joint probability of a new arrival and the probability of finding the desired wavelength to be busy which is given by $1 - p_a + \frac{np_a}{w}$. The set of steady-state equations can be written as

$$P(s_{i_n,i_{n-1},\ldots,i_2,i_1}) = \left[1 - p_a + \frac{np_a}{w} \right]$$
$$\times \left(P(s_{i_n-1,i_{n-1}-1,\ldots,i_1-1}) + P(s_{l,i_n-1,i_{n-1}-1,\ldots,i_1-1}) \right). \quad (3.36)$$

The set of equations in Eqs. 3.35 and 3.36 can be solved by equating the state probabilities as

$$P(s_{i_n, i_{n-1}, \ldots, i_2, i_1}) = P(s_n) \tag{3.37}$$

for any $i_1, i_2, \ldots, i_n \in \{1, 2, \ldots, l\}$ with $i_1 < i_2 <, \ldots, < i_n$ for $n \in \{1, 2, \ldots, \min\{l, w\}\}$.

To obtain $P(s_n)$, Eq. 3.35 can be substituted in Eq. 3.36 to get

$$P(s_n) = \left[p_a - \frac{(n-1)p_a}{w} \right] (P(s_{n-1}) + P(s_n))$$

$$\Rightarrow P(s_n) = \frac{p_a - \dfrac{p_a}{w}.(n-1)}{1 - p_a + \dfrac{p_a}{w}.(n-1)} P(s_{n-1})$$

$$P(s_n) = \left[1 - p_a + \frac{np_a}{w} \right] (P(s_{n+1}) + P(s_n))$$

$$\Rightarrow P(s_{n+1}) = \frac{p_a - \dfrac{p_a}{w}.n}{1 - p_a + \dfrac{p_a}{w}.n} P(s_n). \tag{3.38}$$

Note that both the equations above are similar and consistent in the solution obtained. Using induction on one of the equations gives the general term as

$$P(s_k) = \prod_{i=0}^{k-1} \frac{w - i}{w.\dfrac{1 - p_a}{p_a} + i}.P(s_0), \tag{3.39}$$

where the probability of the initial state can be determined by using the fact that the sum of all the probabilities is unity. That is,

$$P(s_0) + \sum_{n=1}^{\min\{l, w\}} \binom{l}{n} P(s_n) = 1$$

$$\Rightarrow P(s_0) = \left[1 + \sum_{n=1}^{\min\{l, w\}} \binom{l}{n} \prod_{i=0}^{n-1} \frac{w - i}{w.\dfrac{1 - p_a}{p_a} + i} \right]^{-1}. \tag{3.40}$$

Therefore, for any $k \in \{1, 2, \ldots, \min\{l, w\}\}$

$$P(s_k) = \frac{\prod_{i=0}^{k-1} \dfrac{w-i}{w.\dfrac{1-p_a}{p_a}+i}}{1 + \sum_{n=1}^{\min\{l,w\}} \binom{l}{n} \prod_{i=0}^{n-1} \dfrac{w-i}{w.\dfrac{1-p_a}{p_a}+i}}. \qquad (3.41)$$

The steady-state throughput for the general case can be obtained from Eq. 3.32 and using the recursive expressions for the state probabilities given in Eq. 3.41 as

$$\tau(p_a, w, l) = \sum_{i=1}^{l} P(s_1) + \sum_{i=j+1}^{l}\sum_{j=1}^{l-1} 2P(s_2) + \sum_{i=j+1}^{l}\sum_{j=k+1}^{l-1}\sum_{k=1}^{l-2} 3P(s_3) + \cdots$$

$$+ \sum_{i_l=i_{l-1}+1}^{l}\sum_{i_{l-1}=i_{l-2}+1}^{l-1} \cdots \sum_{i_{l-low+1}=1}^{l-low+1} (l \circ w) P(s_{low})$$

$$= l.P(s_1) + \binom{l}{2}.2P(s_2) + \binom{l}{2}.3P(s_3) + \cdots + \binom{l}{l \circ w}.(l \circ w).P(s_{low})$$

$$= \sum_{k=1}^{low} \binom{l}{k}.kP(s_k), \qquad (3.42)$$

where $l \circ w = \min\{l, w\}$ and the probability of being in a state $P(s_k)$ is defined in Eq. 3.41.

The blocking probability of bursts at an OBS node without wavelength conversion capability and with w wavelengths serving fixed length bursts can be derived considering the probabilities of all the states that arise due to blocking of bursts with their corresponding blocking probabilities. Similar to the simpler case considered earlier, if the system is in a state $S_n = s_{i_n,i_{n-1},\ldots,i_1}$, a new burst arriving in the next slot t_i may be blocked either when $i_n = l$ or when $i_n \neq l$. When $i_n = l$, the new burst is blocked with a probability $(n-1)p_a/w$ and the resulting state would be $s_{i_{n-1}+1,i_{n-1},\ldots,i_1+1}$. However, if $i_n \neq l$, the new arrival is blocked with a probability np_a/w and the new state would be $s_{i_n+1,i_{n-1}+1,\ldots,i_1+1}$. Therefore, the blocking probability of the node can be written as

$$PB(p_a, w, l, n) = P(s_n).\left[(n-1)\frac{p_a}{w}.\binom{l-1}{n-1} + n\frac{p_a}{w}.\binom{l-1}{n}\right]$$

$$= P(s_n).\frac{p_a}{w}.\frac{l-1}{l}.n\binom{l}{n}, \qquad (3.43)$$

where the first term is the product of the probability of the blocking when $i_n = l$ and the number of times $i_n = l$ occurs in $s_{i_n,i_{n-1},\ldots,i_1}$ while the second term is computed similarly for the case of $i_n \neq l$. The total blocking probability is simply the blocking probability added over all the states which is

$$PB(p_a, w, l) = \sum_{k=1}^{low} PB(p_a, w, l, n) = \frac{p_a}{w} \cdot \frac{l-1}{l} \cdot \sum_{k=1}^{low} \binom{l}{k} . k P(s_k)$$

$$= \frac{p_a(l-1)}{wl} . \tau(p_a, w, l). \qquad (3.44)$$

3.5.2 Model for Node with Wavelength Conversion

This section computes the steady-state throughput and blocking probability of bursts at an OBS node with wavelength conversion capability. The state model presented here to compute the blocking probability of a node can also be used to consider the case of partial wavelength conversion. In general, let there be w wavelengths and let there be $k \leq w$ wavelength converters so that traffic on only k wavelengths can be converted. The wavelength conversion capability is measured by the factor $c = k/w$ with $c = 0$ indicating no-wavelength conversion and $c = 1$ indicating full wavelength conversion. Unlike the case of no-wavelength conversion, a burst requesting a busy wavelength is transmitted on another wavelength till all the wavelengths are busy or all the k converters are used.

To arrive at the steady-state equations for this case, consider the following cases which lead to the state $S_n = s_{i_n, i_{n-1}, \ldots, i_1}$, where $n \in \{1, 2, \ldots, l \circ w\}$ and $i_1 < i_2 < \ldots < i_n$:

- $i_1 \neq 1$, ($w < l$ and $n = w$): This happens when there is no new burst arrival or the arriving burst is blocked. In this case, the previous state is denoted by $S_w = s_{i_w-1, i_{w-1}-1, \ldots, i_1-1}$ and the transition probability is unity so that the set of steady-state equations is given by

$$P(s_{i_w, i_{w-1}, \ldots, i_1}) = P(s_{i_w-1, i_{w-1}-1, \ldots, i_1-1}). \qquad (3.45)$$

- $i_1 \neq 1$, ($w \geq l$ or $n \neq w$): This state occurs either when there is no new burst or if the new burst requests a wavelength being used and conversion is not possible. The probability of reaching this state can be computed from the probability that the new burst requests for a used wavelength which cannot be converted to another one (due to lack of converters). The previous states for this are either of the type $S_n = s_{i_n-1, i_{n-1}-1, \ldots, i_1-1}$ or of the type $S_{n+1} = s_{l, i_n-1, i_{n-1}-1, \ldots, i_2-1}$. These transitions are shown in Fig. 3.7 with the corresponding transition probabilities. The set of steady-state equations with the transition probabilities can be written as

$$P(s_{i_n, i_{n-1}, \ldots, i_2, i_1}) = \left(P(s_{i_n-1, i_{n-1}-1, \ldots, i_1-1}) + P(s_{l, i_n-1, i_{n-1}-1, \ldots, i_2-1}) \right)$$
$$\times \left[(1 - p_a) + p_a \cdot \frac{n}{w} \cdot (1 - c) \right]. \quad (3.46)$$

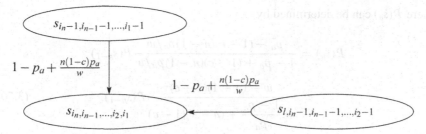

Fig. 3.7 Generic representation of nth state for an OBS node with wavelength conversion: $i_1 \neq 1$ and ($w \geq l \geq 1$ or $n \neq w$)

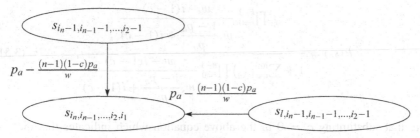

Fig. 3.8 Generic representation of nth state for an OBS node with wavelength conversion: $i_1 = 1$ and ($w \geq l$ or $n \neq w$)

- $i_1 = 1$: This means that a new burst is being served on one of the wavelengths and the previous states are either $S_{n-1} = s_{i_n-1,i_{n-1}-1,\ldots,i_2-1}$ or $S_n = s_{l,i_n-1,i_{n-1}-1,\ldots,i_2-1}$. The state diagram for this case is shown in Fig. 3.8. The probability of this transition includes the probability of a new arrival (p_a) and the probability that the new burst requests an unused wavelength or it can be converted into another unused wavelength. Therefore, the probability of this transition can be written as

$$p_a \left(1 - \frac{n-1}{w} + \frac{n-1}{w}.c \right) = p_a - (1-c)\frac{(n-1)p_a}{w} \qquad (3.47)$$

so that the set of steady-state equations can be written as

$$P(s_{i_n,i_{n-1},\ldots,i_2,i_1}) = \left(P(s_{i_n-1,i_{n-1}-1,\ldots,i_2-1}) + P(s_{l,i_n-1,i_{n-1}-1,\ldots,i_2-1}) \right)$$
$$\times \left[p_a - (1-c)\frac{(n-1)p_a}{w} \right]. \qquad (3.48)$$

The set of steady-state equations can be solved by equating the probability of occurrence for all the S_n states, i.e., for any $i_1, i_2, \ldots, i_n \in \{1, 2, \ldots, l\}$ with $i_1 < i_2 < \ldots < i_n$

$$P(s_{i_n,i_{n-1},\ldots,i_2,i_1}) = P(s_n), \qquad (3.49)$$

where $P(s_n)$ can be determined by

$$P(s_n) = \frac{p_a - (1-c)(n-1)p_a/w}{1 - p_a + (1-c)(n-1)p_a/w} P(s_{n-1})$$

$$= \frac{w - (n-1)(1-c)}{w.\dfrac{1-p_a}{p_a} + (n-1)(1-c)} P(s_{n-1}). \qquad (3.50)$$

For any value of $k \in \{1, 2, \ldots, l \circ w\}$, the above expression can be written as

$$P(s_k) = \frac{\prod_{i=0}^{k-1} \dfrac{w - i(1-c)}{w.\dfrac{1-p_a}{p_a} + i(1-c)}}{1 + \sum_{n=1}^{low} \binom{l}{n} \prod_{i=0}^{n-1} \dfrac{w - i(1-c)}{w.\dfrac{1-p_a}{p_a} + i(1-c)}}. \qquad (3.51)$$

Note that substituting $c = 0$ in the above equation which indicates that there is no-wavelength conversion leads to Eq. 3.41.

The steady-state throughput of the node remains same as that for a node without wavelength conversion, i.e., Eq. 3.42. The blocking probability of the node is also similar to that in the case of no-wavelength conversion except for the additional term of $(1-c)$. Therefore,

$$PB(p_a, w, l, c) = \frac{p_a(l-1)}{wl}(1-c).\tau(p_a, w, l, c), \quad \text{if } w \geq l. \qquad (3.52)$$

When $w < l$, three cases may arise which are given below:

- If $i_n \neq l$ and $n = w$, a new burst is blocked with a probability of one as mentioned earlier. The probability of this event occurring is simply the probability of a new burst arrival p_a. The node enters the state $s_{i_n+1,i_{n-1}+1,\ldots,i_1+1}$.
- If $i_n \neq l$ and $n \neq w$, the node enters the state $s_{i_n+1,i_{n-1}+1,\ldots,i_1+1}$ on a new burst arrival which is blocked with a probability $np_a(1-c)/w$.
- The new burst arrived is blocked with a probability $(n-1)(1-c)p_a/w$ and the state changes to $s_{i_{n-1}+1,\ldots,i_1+1}$ if $i_n = l$.

Considering all the three cases above and following a reasoning similar to that used in the case of no-wavelength conversion, if $n \neq w$, the blocking probability of a node can be written as

$$PB(p_a, w, l, c, n) = \frac{(l-1)p_a(1-c)}{wl}.n\binom{l}{n}P(s_n) \qquad (3.53)$$

and if $n = w$,

$$PB(p_a, w, l, c) = P(s_n) \cdot \frac{p_a}{w} \left[(w-1)(1-c) \binom{l-1}{w-1} + w \binom{l-1}{w} \right]$$

$$= \frac{(l-1)p_a}{wl}(1-c) \cdot w \binom{l}{w} P(s_n) + p_a \binom{l-1}{w} c P(s_n). \quad (3.54)$$

Combining the cases of $n = w$ and $n \neq w$ and summing over all the values of n give the total blocking probability of a node with wavelength conversion to be

$$PB(p_a, w, l, c) = \sum_{k=1}^{w} PB(p_a, w, l, c, k)$$

$$= \frac{p_a(1-c)(l-1)}{wl} \cdot \tau(p_a, w, l, c) + \binom{l-1}{w} p_a c P(s_w). \quad (3.55)$$

The model for blocking probability derived in this section is useful to evaluate the impact of the number of wavelength converters at a node. The model can be used to study if increasing the number of wavelengths gives the same benefit obtained with using wavelength conversion [31]. It also helps to study the effect of burst length (measured in terms of l) on the blocking probability. It was shown in [31] with simple numerical results that when the traffic is low, node with more converters has a lower blocking than that with fewer converters. This advantage is absent when the traffic is high and there is greater benefit in increasing the number of wavelengths. This is because when the traffic is high, there would not be many free wavelengths to get benefited by wavelength conversion capability at a node.

3.6 Blocking Probability with Partial Wavelength Conversion

In full wavelength conversion, a burst arriving on a particular wavelength can be switched to any other wavelength on the outgoing link. Although, full wavelength conversion reduces the blocking probability by a large extent compared to no-wavelength conversion, implementing full wavelength conversion is difficult due to the cost of equipment and technological limitations [22]. It is also known that the output power deteriorates strongly as a function of the distance (number of wavelengths) between the input and output wavelengths [10]. One method to reduce the cost of wavelength conversion at a node is to share a limited number of tunable wavelength converters among multiple links. A few wavelength converters are shared among all the ports either in a share-per-node or share-per-link architecture [11]. In the former, the wavelength converters are placed in a pool at a node and the traffic on any incoming link can use the converter if available. In the latter, a wavelength converter is associated with an output link which can be used by traffic directed only to that link. This kind of wavelength conversion where a few number of full wavelength converters are shared among multiple ports is known as partial wavelength conversion. Though not as effective as full wavelength conversion, partial wavelength conversion reduces blocking probability compared to the case of no-wavelength conversion, while limiting the cost of wavelength converter equipment at a node. Simulations as well as analysis showed that partial wavelength conversion reduces the BLP in OBS networks [11, 12]. The number of such converters required for a desired BLP can also be obtained using the results in [11, 12, 24]. In this section, the BLP at an OBS node with partial wavelength conversion is analyzed using a CTMC model [1, 2]. The models for OBS node are significantly different from those proposed for optical packet or circuit switching networks due to the asynchronous nature of the OBS network [11].

Assume that an OBS node has N input links with W wavelengths in each link and there are $C < W$ wavelength converters at each output link. Assume that the bursts destined to an output link are uniformly distributed across all the links. Bursts arrive following a Poisson distribution with mean λ and have an exponential service time with mean $1/\mu$. Each burst is sent to the output link either with or without wavelength conversion depending on the availability of the wavelength for the duration of burst. The burst is blocked only if either no wavelength converter is available or all the wavelengths are busy. Since it is assumed that bursts are distributed uniformly across all the output links, the BLP is evaluated for a single output link which can be added over all the links to obtain the BLP at a node. Consider the state of a node described by two variables $w(t)$ and $c(t)$, which denote the number of wavelengths and the number of wavelength converters in use at time t, respectively. Let $X(t) = \{w(t), c(t) : t \geq 0\}$ denote a stochastic process on the state space $S = \{(w, c) : 0 \leq w \leq W, 0 \leq c \leq \min\{w, C\}\}$. If the system is in state $S = (w, c)$ at time t and there is a burst arrival in the interval $(t, t + \delta)$, the following transitions are possible:

1. The wavelength requested by the burst is available on the output link with a probability $(W - w)/W$ so that the state changes to $(w + 1, c)$.
2. The requested wavelength is being used with a probability of w/W so that
   - if $c = C$, the burst would be blocked due to unavailability of wavelength converters and the state remains the same or
   - if $c < C$, the burst is scheduled on a wavelength selected randomly and one of the free converters is used. Then the state changes to $(w + 1, c + 1)$.
3. If all the wavelengths are busy so that the state is (W, c) for some c at time t, then a new burst is blocked.

Similarly, when the system is in state (w, c) and a burst is serviced in the interval $(t, t + \delta)$ with a probability $w\mu\delta$, there are two possible states as follows:

1. The new state is $(w - 1, c - 1)$ with a probability c/w if a wavelength converter is used by the burst.
2. The state is $(w - 1, c)$ with a probability $(w - c)/w$ if wavelength converter is not used by the burst.

With the above description of the stochastic process $X(t)$, it is clear that the process is a CTMC. Let the state space be composed of subsets, termed as *levels*, such that within a level w is constant. The state space can be written in terms of levels as

$$S = \{\underbrace{(0,0)}_{\text{level }0}, \underbrace{(1,0), (1,1)}_{\text{level }1}, \underbrace{(2,0), (2,1), (2,2)}_{\text{level }2}, \ldots, \underbrace{(W,0), (W,1), \ldots, (W,C)}_{\text{level }W}\}.$$

(3.56)

Note that the state transitions occur either within a level or across two neighboring levels. Based on all the possible state transitions described earlier and the representation of the state space in the form of levels, the infinitesimal generator Q of the CTMC can be written in a block-tridiagonal form as [2]

$$Q = \begin{bmatrix} L_0 & N_1 & & & & \\ M_0 & L_1 & N_2 & & & \\ & M_1 & L_2 & \ddots & & \\ & & \ddots & \ddots & N_W & \\ & & & M_{W-1} & L_W \end{bmatrix}_{(C+1)(W-\frac{C}{2}+1)},$$

(3.57)

where $L_w = -(\lambda + w\mu)I_{w+1}$ for $w < C$, $L_i = -(\lambda + w\mu)I_{C+1} + \lambda \bar{I}_w$ for $C \leq w < W$, and $L_W = -W\mu I_{C+1}$. Here, I_w denotes an identity matrix of size w and

$$\bar{I}_w = \begin{bmatrix} 0 & 0 & \ldots & 0 \\ 0 & 0 & \ldots & 0 \\ \vdots & \vdots & \ddots & \vdots \\ 0 & 0 & \ldots & \frac{w}{W} \end{bmatrix}_{C+1 \times C+1}.$$

(3.58)

M_{w-1} is μ times the upper-left $(C+1) \times (C+1)$ (upper-left $w+1 \times w$) block of the matrix A_w (which is defined below) if $w > C$ (if $1 \leq w \leq C$). The matrix N_w is λ times the upper-left $(C+1) \times (C+1)$ (upper-left $w+1 \times w+2$) block of B_w (also defined below) if $w \geq C$ (if $1 \leq w < C$). The two matrices A and B are defined as

$$A_w = \begin{bmatrix} w & & & \\ 1 & w-1 & & \\ & 2 & w-2 & \\ & & \ddots & \ddots \end{bmatrix}$$

$$B_w = \begin{bmatrix} \frac{W-w}{W} & \frac{w}{W} & & \\ & \frac{W-w}{W} & \frac{w}{W} & \\ & & \frac{W-w}{W} & \ddots \\ & & & \ddots \end{bmatrix}$$

(3.59)

The steady-state solution for an irreducible and aperiodic CTMC can be found by solving the set of equations with the infinitesimal generator given below [20]

$$xQ = 0$$
$$xv = 1, \tag{3.60}$$

where v is a column of vector of suitable size with all ones. Obtaining stationary solution to the above set of equations using conventional methods of solving Markov chains (for example, the direct methods in [33]) is difficult for a large number of wavelengths (typically above 128 wavelengths per link).

To avoid the complexity in solving the system of equations for a large number of wavelengths, the block-tridiagonal structure of Q can be used for advantage. Although the analysis in [24] is similar to the one given here, it does not take advantage of the block-tridiagonal nature of the infinitesimal generator and hence obtaining the solution is complex. Observe that one of the equations in the first set of equations in Eq. 3.60 is redundant. Let P be a matrix obtained by replacing the entries in the first column of Q by setting $L_0 = \lambda$ and $M_0 = [0 \ 0]^T$. Note that this preserves the block-tridiagonal structure of Q yet giving a nonsingular system of equations. Let $b = [1 \ 0 \ 0 \ \ldots \ 0]$ so that the system of equations can be written as

$$zP = b \tag{3.61}$$

and the steady-state probabilities are given by

$$x = \frac{z}{ze}, \tag{3.62}$$

where e is a column vector of ones. To solve Eq. 3.61, the LU factorization algorithm proposed in [13] is used. The numerical stability of this method for the system described here is discussed in [2] along with the assumptions required for making the solution tractable even for large systems. Finally, in terms of the solution vector x, the BLP is written as

$$PB = x_W e + \sum_{w=C}^{W-1} \frac{w}{W} x_{w,C}, \tag{3.63}$$

where x_w is the solution vector for level w partitioned as $x_w = (x_{w,0}, x_{w,1}, \ldots, x_{w,C})$ for $w \geq C$.

The advantage of the CTMC model proposed for an OBS node with partial wavelength conversion is that it can be used for any general burst length distribution and arrival process. For any different burst arrival process, by suitable definition of the state space, the infinitesimal generator of the Markov chain is obtained. Exploiting the block-tridiagonal property of the generator, the steady-state probabilities can be derived from which the BLP can be computed. The analysis is also extended for a general Markovian arrival process in [2]. The model is useful for evaluating blocking probabilities for large systems and rare blocking probabilities (for example, systems with 256 wavelengths per fiber and blocking probabilities in the order of 10^{-40}) and for provisioning wavelengths and converters for a given traffic distribution.

3.7   Blocking Probability with Limited Wavelength Conversion

As discussed in the earlier section, wavelength converters are costly devices and bear some side effects that degrade the signal quality. Since the output power deteriorates as a function of the distance of conversion, which is the number of wavelengths (or the bandwidth) between the input and the output wavelengths, limited wavelength conversion has gained significant attention in research [10]. In limited wavelength conversion, an incoming wavelength is allowed to be converted to a wavelength only in a subset of wavelengths, termed as the *target range* of the wavelength. Restricting the range of conversion improves the signal quality compared to allowing conversion across all the wavelengths [36]. Computing blocking probabilities with limited wavelength conversion and comparison of its performance with full wavelength conversion have been done for circuit-switched networks in [36]. The analysis assumed link independence where the blocking probability of each link is independent of the other. The assumption has been relaxed for regular topologies by using a probabilistic model in [32]. The reduced-load approximation developed for the case of no-wavelength conversion in [6] was extended using a graph-theoretic model to estimate the path blocking probabilities in a network of arbitrary topology with limited wavelength conversion in [34].

The models for evaluating blocking probability with limited wavelength conversion in circuit-switched networks are not directly applicable to OBS networks due to the unacknowledged bufferless nature of these networks. In this section, the effect of limited wavelength conversion on the blocking probability is discussed with a reduced-load fixed point approximation method popularized by [19] and then with a more generic CTMC model. The overflow fixed point (OFP) model presented here assumes that blocking in each link evolves independently of the blocking in other links allowing the decoupling of analysis [29, 39]. The evolution of blocking probability for a set of wavelengths that belong to a conversion range is represented by a set of fixed point equations which are solved efficiently with successive substitution to obtain the link blocking probabilities. This method is similar to the reduced-load Erlang fixed point method used to obtain the blocking probabilities of links in Section 3.8. The path blocking probabilities are obtained from the link blocking probabilities using a successive substitution algorithm. The OFP method gives the flexibility for use with any wavelength assignment algorithm for any range of wavelength conversion.

3.7.1   Overflow Fixed Point Method

Assume that the burst arrivals on a link, which consist of bursts that arrive freshly from an ingress node and those from the adjacent links, follow a Poisson distribution. Let the burst lengths be iid random variables with a constant mean. The load from adjacent links on a wavelength $i \in \{1, 2, \ldots, w\}$ is denoted by ρ_i while the external load on a link from the ingress node is denoted by ρ_e. Assume that the effect of offset time is negligible so that the load is same either for JET or for JIT reservation scheme. A burst arriving on a wavelength i can be converted only to a wavelength j in the set R_i defined by

$$R_i = \{j \,|\, j = (i + k) \bmod(w), k = 0, \pm 1, \pm 2, \ldots, \pm d\}, \qquad (3.64)$$

where d is an integer specifying the range of conversion assuming modulo $w + 1 \equiv 1$ arithmetic is used [32, 34]. The set R_i is the target range of wavelength i which indicates the possible set of wavelengths over which the bursts on i can be converted. The conversion range is usually specified by the number of wavelengths across which conversion is possible, denoted by $r = 2d + 1$ for $d \in [0, \lfloor w/2 \rfloor]$. The modulo arithmetic is a simplification used for the sake of analytical tractability and neglects the edge effect due to wraparound of wavelength spectrum assumed at the edges [32, 34]. Among the wavelengths in the set R_i, the actual wavelength assigned depends on the wavelength assignment policy used.

In this analysis two such policies are used although this can be extended to any other policy [29]. In a random wavelength assignment policy (RdWA), the wavelengths in R_i are ordered randomly and the first free wavelength is assigned on the output link. The nearest-wavelength-first (NWF) policy tries to minimize the deterioration of the signal quality by choosing a wavelength in R_i that has minimum distance d to the incoming wavelength. Any tie is broken arbitrarily. A burst is blocked if all the output wavelengths in R_i of the input wavelength are busy. The policy for wavelength assignment is only for a burst arriving from an adjacent link while an external burst from an ingress node can be scheduled to any of the free wavelengths in the target range. For analytical tractability, an external burst is assumed to be assigned to any free wavelength in R_i with a probability p_i independent of the distribution of the free wavelengths such that $\sum_{i=1}^{w} p_i = 1$. Therefore, the offered load on a target range of wavelengths R_i due to external bursts is given by $\rho_i + p_i \rho_e$.

Let $N_i(t)$ be the number of bursts scheduled on a wavelength i at time t so that the process $\mathbf{N}(t) = (N_1(t), N_2(t), \ldots, N_w(t))$ is Markovian with $(1 + F)^w$ states where F is the number of fibers in a link. Computation of the stationary distribution of $\mathbf{N}(t)$ is intensive due to the large number of wavelengths and fibers as well as the fact that the evolution of $N_i(t)$ and $N_j(t)$ for any two wavelengths $i \neq j$ is not independent. To make this computation tractable, consider the evolution of the process $\mathbf{N}(t)$ for a target range of a wavelength, i.e.,

$$\mathbf{N}^i(t) = \{N_k(t) : k \in R_i\}, \qquad i \in \{1, 2, \ldots, w\} \qquad (3.65)$$

and assume that the evolution of $\mathbf{N}^i(t)$ is independent of $\mathbf{N}^j(t)$ for $i \neq j$. Note that the process $\mathbf{N}^i(t)$ has only $(1 + F)^{2d+1}$ states and the stationary distribution can be obtained efficiently by solving the balance equations arising from the transition rates described below. Note that each range of wavelengths R_i, $\forall i$ on a link receives bursts from adjacent links and external bursts on i which may be assigned to any wavelength on R_i. In addition to that there are bursts that overflow from adjacent ranges. It is assumed that the conversion range r is much smaller than the number of wavelengths w. Due to this, there is an overflow of bursts from one range of wavelengths to another one. The overflow load on a wavelength $j \in R_i$ is denoted by $\rho_i'(j)$. Since j might be in two ranges R_i and R_k, the evolution of $\rho_i'(j)$ and $\rho_k'(j)$ is not independent for $i \neq k$.

Given a state $\mathbf{N}^i(t) = \mathbf{x}$ such that $N_j(t) = x_j$, let $F_i(\mathbf{x}), i \in \{1, 2, \ldots, w\}$ be the set of outgoing wavelengths that can be selected by the NWF policy which can either be a singleton or doubleton set. Let $n_i(\mathbf{x}) = \sum_{j \in R_i} I\{x_j = F\}$, where $I\{x_j = F\} = 1$ iff $x_j = F$, otherwise $I\{x_j = F\} = 0$. The condition $x_j = F$ indicates that wavelength j is busy on all the fibers. Given the state $\mathbf{N}^i(t) = \mathbf{x}$, such that $N_j(t) = x_j > 0$, i.e., j is busy on at least one of the fibers, the transition rate to state $\mathbf{N}^i(t + dt) = \mathbf{x} - \mathbf{u}_j$, where $\mathbf{u}_j$ is a vector of size $2d + 1$ with jth element equal to 1 while the others are 0, is simply given by x_j because of normalized burst transmission rate and the insensitivity of Erlang blocking formula to the distribution of service time.

Similarly, given a state $\mathbf{N}^i(t) = \mathbf{x}$ such that $N_j(t) = x_j < F$, i.e., wavelength j is free at least on one of the fibers, the transition rate to a state $\mathbf{N}^i(t + dt) = \mathbf{x} + \mathbf{u}_j$ is given as follows for the two wavelength assignment policies:
for the RdWA policy

$$\rho_i'(j) + \frac{\rho_i + \rho_i \rho_e}{2d + 1 - n_i(\mathbf{x})} \tag{3.66}$$

and for the NWF policy

$$\rho_i'(j) + \frac{\rho_i \rho_e}{2d + 1 - n_i(\mathbf{x})} + \frac{\rho_i}{|F_i(\mathbf{x})|}, \quad j \in F_i(\mathbf{x}),$$

$$\rho_i'(j) + \frac{\rho_i \rho_e}{2d + 1 - n_i(\mathbf{x})}, \quad j \notin F_i(\mathbf{x}). \tag{3.67}$$

Note that $n_i(\mathbf{x}) \neq 2d + 1$ since the wavelength j is free at least on one fiber. Once the transition rates are defined, the stationary distribution of the process $\mathbf{N}^i(t)$ can be obtained by solving the balance equations. Given a wavelength assignment (conversion) policy $P \in \{RdWA, NWF\}$, the stationary probability of being in state $\mathbf{N}^i(t) = \mathbf{x}$, denoted by $\pi_i^P(\mathbf{x})$, can be obtained. The blocking probability for an external burst is then estimated as $\sum_{i=1}^{w} p_i \pi_i^P(F, F, \ldots, F)$ and for the burst from an adjacent link as $\pi_i^P(F, F, \ldots, F)$.

The next part in determining the transition rates given by Eqs. 3.66 and 3.67 is to determine the overflow load $\{\rho_i'(j)\}$. For a given wavelength conversion policy, the overflow load is related to the stationary probabilities. For a given $P \in \{RdWA, NWF\}$, with a load $\rho_i'(j), i, j \in \{1, 2, \ldots, w\}$, it is enough the compute the set of probabilities $\{\pi_k^P : |k - i| \leq 2d, i \neq k, |k - j| \leq d\}$.

For the RdWA policy. Given a state $\mathbf{N}^i(t) = \mathbf{x}$, such that $N_j(t) = x_j < F$, a burst arriving from an adjacent link on a wavelength k is assigned to another wavelength $j \in R_j$ with a probability $1/2d + 1 - n_k(\mathbf{x})$ and overflows to an outgoing wavelength $l \in R_i$ with the same probability if $|k - i| \leq 2d, i \neq k$, and $|k - j| \leq d$. Otherwise, the overflow probability is zero. If $\mathbf{N}^k(t) = \mathbf{x}$, such that $N_j(t) = x_j = F$, there is no overflow to wavelength j since it is busy on all the fibers. Therefore, the overflow load to a wavelength $j \in R_i$ can be obtained by adding the overflow from all incoming wavelengths k over the set of states where at least one of the fibers has a free wavelength j. This is written as

$$\rho_i'(j) = \sum_{k:|k-j|\leq d, i\neq k} (\rho_k + p_k \rho_e) \sum_{\mathbf{x}:x_j < F} \pi_k^{RdWA}(\mathbf{x}:x_j < F)/(2d + 1 - n_k(\mathbf{x})),$$
$$\forall i, j \in \{1, 2, \ldots, w\},$$
$$(3.68)$$

where $\pi_k^{RdWA}(\mathbf{x}:x_j < F)$ is the conditional stationary distribution of being in state $\mathbf{N}^k(t) = \mathbf{x}$ given that $N_j(t) = x_j < F$.

For the NWF policy. Given a state $\mathbf{N}^i(t) = \mathbf{x}$, such that $N_j(t) = x_j < F$, a burst arriving from adjacent link on wavelength k overflows to $j \in R_i$ with probability 1 if $|F_k(\mathbf{x})| = 1$, i.e., $F_k(\mathbf{x}) = \{j\}$ and with a probability $1/2$ if $|F_i(\mathbf{x})| = 2$ and $j \in F_k(\mathbf{x}), |k - i| \leq 2d, i \neq k$ and $|k - j| \leq d$. If $j \notin F_k(\mathbf{x})$ either because $F_k(\mathbf{x}) = \{\phi\}$ or if $\nexists j \in F_k(\mathbf{x}), |k - i| \leq 2d, i \neq k$ and $|k - j| \leq d$, there is no overflow of bursts to a wavelength j. Considering all the conditions of overflow, let $S_k(j)$ and $T_k(j)$ be the sets of states in which a burst from an adjacent link on wavelength k is assigned to wavelength j with probability 1 and 1/2, respectively, and let $U_k(j) = S_k(j) \cup T_k(j)$. The overflow load for this case can be obtained by adding all load over all the incoming wavelengths k being converted to $j \in R_i$ and adding this over all the sets of states $\mathbf{x}$ of each target range for which $j \in F_k(\mathbf{x})$, which is mathematically written as

$$\rho_i'(j) =$$

$$\sum_{k:|k-j|\leq d, i\neq k} \rho_k \left(\sum_{\mathbf{x}\in S_k(j)} \pi_k^{NWF}(\mathbf{x}:\mathbf{x} \in U_k(j)) + \sum_{\mathbf{x}\in T_k(j)} \pi_k^{NWF}(\mathbf{x}:\mathbf{x} \in U_k(j))/2 \right)$$
$$\forall i, j \in \{1, 2, \ldots, w\}, \quad (3.69)$$

where $\pi_k^{NWF}(\mathbf{x}:\mathbf{x} \in U_k(j))$ is the conditional stationary distribution of being in state $\mathbf{N}^k(t) = \mathbf{x}$ given that $\mathbf{x} \in U_k(j)$. Note that even when NWF policy is used, for the external bursts only RdWA policy is used so that the overflow load due to the external bursts is given by Eq. 3.68 with $\rho_k = 0$.

Observe that the computation of overflow load and the stationary distribution of state are coupled. Given the load due to bursts from adjacent links on a wavelength $i \in \{1, 2, \ldots, w\}$, ρ_i and that due to external bursts $p_i \rho_e$ and a wavelength conversion policy $P \in \{RdWA, NWF\}$, the relation between the overflow load $\rho'_i(j), i, j \in \{1, 2, \ldots, w\}$ and the stationary distribution $\{\pi^P_k : |k - j| \leq d, i \neq k\}$ is given through a set of fixed point equations which can be solved by a successive substitution algorithm. The procedure is started by initializing $\rho'_i(j)$ and π^P_k to an arbitrary value and for each iteration compute the stationary distribution by solving the balance equations arising from the definition of transition rates in Eq. 3.67 or 3.66 with the overflow load from the previous iteration. This procedure is repeated till the stationary probabilities converge within a predefined limit. The new stationary distributions are used in computation of the overflow load using Eq. 3.68 or 3.69 and the iterations are continued till convergence is arrived in the value of the stationary distributions.

Once the link blocking probabilities are computed, the path blocking probabilities are obtained by using the reduced-load approximation method which accounts for the blocked load along the links on the path before the current link. For each path, the reduction in the load due to the bursts blocked ahead in the routes traversing the path is computed which is used in the computation of the new blocking probabilities. Through the successive substitution method described earlier, a set of fixed point equations that relate the blocking probability to the reduced load along the path are solved to obtain a converged set of blocking probabilities for each path in the network. Even though the existence of fixed point solution is not guaranteed, it has been shown that for different initializations the successive substitution procedure converges to the same fixed point within a few iterations [29]. Simulations and analysis show that compared to the case of no-wavelength conversion, limited wavelength conversion even with a smaller range reduces the average path blocking probability by several orders of magnitude particularly at low loads [29]. The benefit of limited wavelength conversion reduces as the load increases. However, unlike the results in circuit switching that show that even a small range of conversion gives similar performance as that of full wavelength conversion, in OBS networks the performance of limited range conversion is not equivalent to that of full wavelength conversion even with a large range. The analysis presented in this section helps in dimensioning the number of wavelengths or the conversion range required for a given blocking probability.

The OFP method assumes that the conversion range is much smaller than the number of wavelengths. The complexity of the method increases with the number of wavelengths due to the iterative nature of the model. Approximate solutions for a large number of wavelengths with large conversion range have been proposed in [26]. However, the blocking probabilities have been obtained for ON–OFF sources and are not exact for large wavelengths. A single OBS switch has been modeled for small number of wavelengths with $d = 1, 2$ which cannot be generalized easily. The OFP method not only models the blocking probability of a single switch but also uses iterative method to compute the path blocking probabilities. In the next section an OBS switch with limited wavelength conversion is modeled with a generic birth–death process with discouraged arrivals [23]. This queueing model can compute the blocking probability of a switch with any number of wavelengths and for any conversion range.

3.7.2 Queueing Model

Assume that bursts may come at an output port from any input port on any wavelength with uniform probability. Those bursts that arrive on a wavelength i compete for a wavelength j in the set R_i. Assume that bursts on a single wavelength i destined to a set R_i arrive following a Poisson process with a mean arrival rate of λ so that the total arrival rate on all the wavelengths is $w\lambda$. Assuming that the arrival on each wavelength is independent, the arrival process can be assumed to be consisting of $n_i, i = 0, 1, 2, \ldots, w - 1$ independent sources. Assume that each independent source generates bursts with exponentially distributed inter-arrival times of mean $1/\lambda$ and exponentially distributed length with mean $1/\mu$ so that the traffic intensity at a port is $\rho = w\lambda/\mu$. Following the setting in the previous section, a burst from any source n_i is dropped only if the set of wavelengths in the conversion range R_i is busy.

Let the state of the system at time t be described by a stochastic process $\{S(t) = k\}, k = 0, 1, 2, \ldots, w$ if k wavelengths are occupied. The process $S = \{S(t)\}$ is a birth–death process with $w + 1$ states as shown in Fig. 3.9 [23]. In each state k, the arrival rate is denoted by $\lambda_k \leq w\lambda$. Note that the actual arrival rate per output port, λ_k, is not always $w\lambda$ because, when the number of occupied wavelengths is more than the conversion range, some sources of bursts would be blocked. Obviously greater the value of k, lesser would be the value of λ_k. In each state, the service rate is given by $k\mu$. A queueing system with these properties is termed as a *birth–death process with discouraged arrivals* [20].

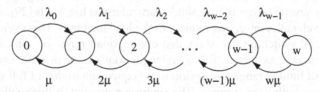

Fig. 3.9 Birth–death process representing the state of a node with limited wavelength conversion

In any state k when the number of bursts $b \geq k$, the average number of bursts blocked, denoted by $n_{b|(k,w)}$, is given by [23]

$$n_{b|(k,w)} = \begin{cases} w\binom{k-b}{w-2-b}/\binom{k}{w}, & \text{for } k = 1, 2, \ldots, w - 2 \\ & \text{and } b = 1, 2, \ldots, k \\ 1, & \text{for } k = w - 1, w \text{ and } b = k \\ 0, & \text{otherwise.} \end{cases} \tag{3.70}$$

The effective arrival rate in each state λ_k is obtained using $n_{b|(k,w)}$. Note that for $k < r$, there is at least one free wavelength in each conversion range R_i so that no burst is blocked and the effective arrival rate is $w\lambda$. For $r \leq k < w$, the number of bursts blocked is obtained by adding the average number of bursts blocked (defined by $n_{b|(k,w)}$) in each state. For $k = w$, all the bursts are blocked so that the effective arrival rate is zero. Thus,

$$
\lambda_k = \begin{cases}
w\lambda, & \text{for } k = 0, 1, \ldots, r - 1 \\
\lambda \left(w - \sum_{b=r}^{k} (b - r + 1) n_{b|(k,w)} \right), & \text{for } k = r, \ldots, w - 1 \\
0, & \text{for } k = w.
\end{cases} \tag{3.71}
$$

With these definitions, it is not difficult to arrive at the equilibrium distribution of the system denoted by π, on solving the set of steady-state balance equations:

$$
\pi_0 = \left[1 + \frac{\lambda_0}{\mu} + \frac{\lambda_0 \lambda_1}{2\mu^2} + \cdots + \frac{\lambda_0 \lambda_1 \ldots \lambda_{w-1}}{w! \mu^w} \right]^{-1} \tag{3.72}
$$

$$
\pi_i = \frac{\lambda_0 \lambda_1 \ldots \lambda_{i-1}}{i! \mu^i} \pi_0, \quad i = 1, 2, \ldots, w. \tag{3.73}
$$

Using the equilibrium distribution and the effective arrival rates in each state, the blocking probability of the system can be obtained as

$$
PB = \sum_{i=0}^{w} \left(\frac{w\lambda - \lambda_i}{w\lambda} \right) \pi_i = \sum_{i=0}^{w} \left(1 - \frac{\lambda_i}{w\lambda} \right) \pi_i. \tag{3.74}
$$

Compared with the OFP model and the approximate model in [26], the queueing model is more generic and scales well for a large number of wavelengths. Simulations in [23] show that the model is exact even for heavy load compared to the OFP model. Although the blocking probability is evaluated for a single OBS node it can be used in any method like the fixed point method to compute the path blocking probabilities of the network.

3.8 Blocking Probability: Erlang Fixed Point Method

Even after the research on effective ways to compute blocking probability in networks similar to OBS networks with features like delayed reservation and preemptive segmentation, exact solutions are found to be unattainable. The difficulties in modeling blocking probability in OBS networks have led researchers to use a single link model (like those in the previous sections) that provides crude upper bounds for the blocking. The single link model does not consider the increased load due to the bursts that are eventually discarded and the reduced load resulting from the blocked bursts. These limitations led to the use of reduced-load fixed point approximation where the link blocking probability is evaluated by considering only the reduced offered load caused by blocking in the other upstream links (links before the one being considered on a path) [27, 28]. The fixed point method helps in the evaluation of blocking probability in different reservation policies and study the effect of burst routing and admission control, and other networking aspects [19].

Let $X_r(t)$ be the number of bursts on route $r \in \mathcal{R}$ at time t, and $Y_r(t)$ be a vector of size $X_r(t)$ in which every element represents the time elapsed from the moment that the corresponding burst has finished its transmission on the first wavelength on its route until time t. Since the transmission times are exponential and subsequent wavelengths are used at fixed offset times, the process $\{X_r(t), Y_r(t) | r \in \mathcal{R}, t \geq 0\}$ is a Markov process. Further $X_r(t)$ is finite and $Y_r(t)$ evolves linearly in time with a bounded interval so that the process has a stationary distribution. Since the exact evaluation of the stationary blocking probability of an arbitrary burst on an arbitrary route is difficult, it is assumed that each blocking event occurs independently between a link and another link along a route. Therefore, the load offered to a link j, ρ_j, satisfies

$$\rho_j = \mu_j^{-1} \sum_{r \in \mathcal{R}} \lambda_r \prod_{i=1}^{J}(1 - I(i, j, r).B_i), \tag{3.75}$$

where $I(i, j, r)$ equals 1 if $(i, j) \in r$ and link i strictly precedes link j on route r in a set of routes $\mathcal{R}$ and B_i is an element of the vector of stationary link blocking probabilities $\mathbf{B} = (B_1, B_2, \ldots, B_J)$ for J directional links in the network. The transmission rate of link j is denoted by μ_j and the bursts arrive on a route r following a Poisson process with a rate λ_r. Following the independence assumption, the individual link blocking probabilities are defined by the Erlang formula in Eq. 3.1. Combining Eqs. 3.1 and 3.75 gives the Erlang fixed point (EFP) equations satisfied by the approximate link blocking probabilities,

$$B_j = E(\mu_j^{-1} \sum_{r \in \mathcal{R}} \lambda_r \prod_{i=1}^{J}(1 - I(i, j, r).B_i), W). \tag{3.76}$$

Resolving the vector $\mathbf{B}$ from the EFP equations in Eq. 3.76 and with the independence assumption, the approximate blocking probability of bursts offered on route r, $B(r)$, satisfies

$$B(r) = 1 - \prod_{i \in r}(1 - B_i) \tag{3.77}$$

and the blocking probability of an arbitrary burst satisfies

$$B = \frac{1}{\Lambda} \sum_{r \in \mathcal{R}} \lambda_r.B(r), \tag{3.78}$$

where $\Lambda = \sum_{r \in \mathcal{R}} \lambda_r$.

To solve the EFP equations in an efficient way, the following successive substitution procedure can be used. Given the vector of blocking probabilities $\mathbf{B}^0 = \mathbf{B}$, let

$T(\mathbf{B})$ be the transformation vector with elements $T_1(\mathbf{B}), \ldots, T_J(\mathbf{B})$ defined by

$$T_j(\mathbf{B}) = E(\mu_j^{-1} \sum_{r \in \mathcal{R}} \lambda_r \prod_{i=1}^{J} (1 - I(i, j, r).B_i), W). \qquad (3.79)$$

Starting with an initial blocking probability vector $\mathbf{B}$ by repeatedly computing $\mathbf{B}^n = T(\mathbf{B}^{n-1})$ for $n = 1, 2, \ldots$ until $\mathbf{B}^n$ is sufficiently close to $\mathbf{B}^{n-1}$, the equations can be solved (if a solution exists). Unlike the application of EFP equations in traditional circuit switching networks, the uniqueness of the solutions cannot be established. But note that the transformation $T(\mathbf{B})$ is a continuous mapping from the compact set $[0, 1]^J$ to itself and therefore has a fixed point according to the Brouwer fixed point theorem [25]. Convergence of these equations has been observed in many numerical simulations done in the literature [27].

The independence assumption helps in decoupling the blocking probabilities among the network links, yet takes into account the reduced offered load resulting from blocking at the upstream nodes along the path. Assume Poisson arrival helps to use the Erlang formula for blocking probability at each link. If any other distribution is assumed for the arrival of bursts, if closed-form expressions can be attained for the blocking probability the Erlang formula can be replaced by the corresponding formula. As long as the transformation vector is decreasing, the convergence of such equations is also guaranteed.

The EFP method of obtaining the blocking probabilities can be applied under both JET and JIT regimes for which the corresponding single link model is given in Section 3.2. With JET signaling, the transmission rate of link j, μ_j, is set to $1/E(S_j)$ where S_j is the burst service time in link j. In JIT signaling it is simply replaced by $1/E(S_j) + E(D_j)$ where D_j is the offset time. Replacing for the service time in Eqs. 3.1 and 3.76 gives the blocking probability in these two regimes.

3.9 Blocking Probability for Multi-class OBS Networks

So far, all the sections discussed models for blocking probability of an OBS node assuming that all the bursts have equal priority and a burst that contends with another burst in transmission is always dropped. But in a network that supports differentiated levels of service or that carries various types of traffic with different requirements, prioritized dropping is required at an OBS node. It is not difficult to support prioritized traffic at the burst level with the network architecture mentioned in Chapter 1. At an ingress node, since each burst is assembled based on the destination egress node and the type of traffic (or QoS requirements), all the bursts belonging to a class can be assembled at the same queue. A simple way of creating priority among the bursts is to give an extra offset time to the bursts that require higher priority [38]. The method of using additional offset time for bursts with higher priority is simple and efficient in OBS networks without FDL buffers. Using extra offset time also avoids the complexity of the FDL-based implementations. Service differentiation with extra offset time has been studied in bufferless OBS networks [37] as well as those with FDL buffers [38]. In this section, the blocking probability of a core node is evaluated with prioritized burst traffic.

As mentioned earlier, if there is only one class of traffic in the network and the remaining offset time at any node is equal for all the bursts traveling on same hop length path, the blocking probability at a core node in OBS network can be approximately characterized by the Erlang B formula for an $M/M/W/W$ queue. Since the Erlang B formula is insensitive to the service time distribution, the blocking probability can be evaluated using the same formula for an $M/G/W/W$ queue also. However, when there is more than one priority class, the application of $M/G/W/W$ queue to evaluate the blocking probability is not straightforward [35]. This is due to the fact that the offset time of every burst in a single-class OBS system is the same and it can be neglected for the purpose of analysis. But the blocking process in multi-class OBS networks is not simple due to the variable offset times of different classes. The phenomenon of retro-blocking described in Chapter 1 also occurs due to the variable offset times given to bursts of different classes.

Due to the effect of retro-blocking, a reservation made earlier in time is blocked by a BHP arriving later with a smaller offset time. When there is such preemption of bursts, the blocking probability of lower priority traffic cannot be directly evaluated from the Erlang B formula applied on the load of the low priority traffic alone. One simple method to evaluate the blocking of each class of traffic when there are only two classes or there are multiple classes of traffic with perfect class isolation is to use the conservation law. This procedure is explained further in this section. In general, if all the classes of traffic have burst sizes of exponential distribution, due to the memoryless property, Erlang B formula can be applied directly with the combined load arising from the traffic of all the classes. If the burst size distribution is not exponential, this method cannot be used except for small blocking probability (or low load) values. To enable the application of classical Markov theory for multi-class network, it is assumed that either there is total or a high degree of isolation among the classes [35, 37] or the OBS network is work conserving, i.e., conservation law holds good [9, 35]. For small blocking probability values, due to the insensitivity of the Erlang B formula to service time distribution as well as the rarity of a preemption event, class isolation can be assumed among traffic from different classes.

One method to enforce strict priority among different classes is to set the offset time of class i, O_i, with a higher priority than the other classes $j, n \geq i > j \geq 1$ to $\sum_{j=1}^{i-1}(L_i + H_i^{\max}\delta) + H_i^{(sd)}\delta$, where δ is the BHP processing time at each node, L_i is the upper bound on the average length of the bursts of class i, $H_i^{(sd)}$ is the random variable indicating the number of hops traveled by bursts of class i for an sd (source–destination) pair and $H_i^{\max} = \max_{(sd)}\{H_i^{(sd)}\}$ [35].

If there are n classes of traffic, let bursts of class i have a higher priority over the bursts of class j for $i > j$. When two bursts from classes i and j contend with each other, bursts from class j are dropped. Assume that bursts from class i arrive at a core node following a Poisson process with mean λ_i and the service time is exponentially distributed with mean $\mu_i = 1/L_i$ where L_i is the mean length of the bursts of class i. The traffic load due to bursts of class i on a wavelength is $\rho_i = \lambda_i/\mu_i$ and the total traffic load is $\rho = \sum_{i=1}^{n} \rho_i$. The next step would be to determine the lower and upper bounds on the BLP for the traffic of a given class i.

To derive the upper bound PB_i^{upp} and lower bound PB_i^{low} for class i traffic, a fundamental assumption to be made is that of traffic isolation. That is, class i traffic is completely isolated from the effects of the traffic due to other classes. Traffic from lower priority classes does not affect the BLP of higher priority classes. This assumption makes it easier to compute the BLP of traffic of class i using the Erlang B formula so that $PB_i = E(\rho_i, W)$ and for traffic of another class j, the BLP can be written as $PB_j = E(\rho_j, W)$ where the load due to class j does not affect the load and loss rate of class i. Due to the assumption of complete isolation among different classes, a conservation law is written for OBS networks [38] which holds good for a large extent with bursts of equal lengths. However, the conservation law does not hold good in case of bursts of unequal average length [40], particularly when the bursts of low-priority class are much longer than the high-priority bursts [4, 9]. Assuming that the conservation law holds good, the overall loss probability due to traffic of all classes can be expressed as a convex combination of the BLP of the individual classes. For class i, if the BLP is PB_i, the overall BLP is

$$PB(\rho, W) = E(\rho_1 + \rho_2 +, \ldots, +\rho_i, W) = \sum_{k=1}^{i} \frac{\lambda_i}{\lambda_1 + \lambda_2 +, \ldots + \lambda_i} PB_i. \quad (3.80)$$

As mentioned in Section 3.2, the upper and lower bounds on the BLP in an OBS network with JET protocol and FDLs is given by Eqs. 3.1 and 3.2, respectively. For traffic with highest priority, say n, it is assumed that the loss is zero since it always wins contention with bursts from other lower priority classes. For the class $n - 1$, the upper and lower bounds on the loss probability are given by $PB(\rho_{n-1}, W)$ and $PB(\rho_{n-1}, W, d)$ computed directly from Eqs. 3.1 and 3.2, respectively.

For the traffic from classes of lower priority, the average loss probability over the traffic from classes $n - 1$ to i, denoted by $PB_{n-1,i}$, is evaluated as $\sum_{k=i}^{n-1} \frac{\rho_k}{\rho} . PB_k$. With these definitions, the upper bound on the BLP for class $n - 2$ can be written as

$$\text{PB}_{n-1,n-2}^u = \sum_{k=n-2}^{n-1} \frac{\rho_k}{\rho} . PB_k^u \quad (3.81)$$

where PB_k^u can be computed from Eq. 3.1. $PB_{n-1,n-2}^u$ can also be computed from Eq. 3.1 by substituting the sum of loads due to class $n - 1$ and $n - 2$ defined by $\rho_{n-1,n-2} = \sum_{k=n-2}^{n-1} \rho_k$. That is,

$$PB_{n-1,n-2}^u = E(\rho_{n-1,n-2}, W). \quad (3.82)$$

Equating both Eqs. 3.81 and 3.82 give

$$PB_{n-2}^u = \frac{E(\rho_{n-1,n-2}, W) - \frac{\rho_{n-1}}{\rho}.PB_{n-1}^u}{\rho_{n-2}/\rho}. \tag{3.83}$$

Similarly, the lower bound computed from Eq. 3.2 can be equated with

$$PB_{n-1,n-2}^l = E(\rho_{n-1,n-2}, W, D) = \sum_{k=n-2}^{n-1} \frac{\rho_k}{\rho}.PB_i^l \tag{3.84}$$

so that the lower bound on the BLP of class $n-2$ can be written as

$$PB_{n-2}^l = \frac{E(\rho_{n-1,n-2}, W, D) - \frac{\rho_{n-1}}{\rho}.PB_{n-1}^l}{\rho_{n-2}/\rho}. \tag{3.85}$$

This procedure can be used to obtain the upper and lower bounds on BLP for any class. In general, for traffic due to class i, $0 \le i \le n-2$, the upper bound can be written as

$$PB_i^u = \frac{E(\rho_{n-2,i}, W) - \sum_{k=i+1}^{n-1} \frac{\rho_k}{\rho}.PB_k^u}{\rho_i \rho} \tag{3.86}$$

and the lower bound can be written as

$$PB_i^l = \frac{E(\rho_{n-2,i}, W, D) - \sum_{k=i+1}^{n-1} \frac{\rho_k}{\rho}.PB_k^l}{\rho_i \rho}. \tag{3.87}$$

In terms of the blocking probabilities, the above equations can be written using the conservation law in Eq. 3.80 as follows. The blocking probability of class i can be calculated as

$$PB_i = \frac{1}{\lambda_i} \left(\sum_{j=1}^i \lambda_j \right) \left[PB_{(1,2,...,i)} - p_{1,2,...,i-1} PB_{(1,2,...,i-1)} \right], \tag{3.88}$$

where

$$p_{1,2,...,i-1} = \frac{\sum_{j=1}^{i-1} \lambda_j}{\sum_{j=1}^i \lambda_j} \tag{3.89}$$

and the total loss probability for all the classes up to i, $PB_{(1,2,...,i)} = E(\sum_{j=1}^i \rho_i, W)$ is evaluated from Eq. 3.1.

The analysis above assumes 100% isolation among the traffic of different classes for which the difference between the offset values of any two classes must be at least five times more than the average length of the bursts of the lower priority class (in general, larger than the maximum length of the low-priority bursts) [38]. However, in practice using large offset times to achieve total isolation among different traffic classes increases the end-to-end delay by a large factor which degrades the delay performance of traffic at the cost of loss performance. Complete isolation by increased offset times may not be practical depending on the traffic characteristics at the edge node, burst assembly mechanisms, and the delay requirements of the high-class traffic.

When complete isolation is not possible, the loss of high-priority bursts is caused by not only the load of high-priority traffic but also a portion of the low-priority traffic being serviced before preemption [9]. For example, when there are two classes of traffic with the load of low-priority class indicated by ρ_l and that of high-priority class by ρ_h, the overall BLP can be obtained using Eq. 3.82. But the BLP of high-priority traffic can be approximated by

$$PB_h = E(\rho_h + \rho_l'(\delta_h - \delta_l), W), \tag{3.90}$$

where $\rho_l'(\delta_h - \delta_l)$ indicates the load due to the low-priority bursts that are already under transmission prior to the arrival of the high-priority bursts and are still under transmission. $\delta_h - \delta_l$ indicates the starting time of the transmission of high-priority burst. $\rho_l'(\delta_h - \delta_l)$ is given by the product of the traffic being carried at the time of the arrival of the high-priority burst and the probability that the low-priority burst has not finished transmission during the period $[t, t + (\delta_h - \delta_l)]$ for some random t. This can be approximated by [9]

$$\rho_l'(\delta_h - \delta_l) = \rho_l(1 - PB_l).(1 - F_l^f(\delta_h - \delta_l)), \tag{3.91}$$

where $\rho_l(1 - PB_l)$ indicates the portion of the traffic of low priority carried by the arrival time of the high-priority burst and $1 - F_l^f(\delta_h - \delta_l)$ is the complementary distribution function of the forward recurrence time [30] of the burst transmission time at time $(\delta_h - \delta_l)$. The forward recurrence time also describes the probability that the low-priority burst which is under transmission from t has not yet finished transmission till $t + (\delta_h - \delta_l)$ [9]. The BLP of the low-priority traffic, PB_l, can be obtained from the conservation law in Eq. 3.80 applied for two classes, i.e.,

$$\rho_h PB_h + \rho_l PB_l = (\rho_h + \rho_l)E(\rho_h + \rho_l, W). \tag{3.92}$$

It can be seen from Eqs. 3.90 and 3.91 that the BLP of high-priority class is dependent on the burst length distribution of the low-priority traffic and not its own. Though the longer bursts are dropped with a higher probability, Eq. 3.91 provides a good approximation for estimating the impact of low-priority traffic on the traffic with a higher priority. An iterative method can be used to solve for PB_l and PB_h since they are coupled.

Even when there are multiple classes of traffic, the BLPs can be calculated in a similar manner as above [8]. Assume that there are n classes of traffic and there is no isolation among the traffic of different classes. Unlike the earlier case of 100%

isolation, it cannot be assumed that the class n bursts are not lost. The BLP of traffic from the nth class can be evaluated by taking into account its own load ρ_n as well as the interfering load due to bursts of lower priority still under transmission at the arrival time. This interfering load for traffic of class n from a lower priority class $i < n$ is denoted by $\rho_i'(\delta_n - \delta_i)$ and the BLP of class n traffic is given by

$$PB_n = E(\rho_n + \sum_{i=1}^{n-1} \rho_i'(\delta_n - \delta_i), W). \qquad (3.93)$$

For any given set $S_k = \{1, 2, \ldots, k \le n - 1\}$ of classes, the conservation law given in Eq. 3.80 holds good, i.e.,

$$(\sum_{i=0}^{k} \rho_i).PB_{S_k} = \sum_{i=0}^{k} \rho_i PB_i. \qquad (3.94)$$

For every class $j \in S_k$, there is interference from all other classes $i \notin S_k$ given by

$$\rho_i'(\delta_j - \delta_i) = \rho_i(1 - PB_i).(1 - F_i^f(\delta_j - \delta_i)), \qquad (3.95)$$

where the terms have same interpretation as in Eq. 3.91. For a given value of k, to compute the BLP of set S_k, Erlang formula is used by including the load from the interfering components multiplied by the relative occurrence of individual classes in S_k which turns out to be

$$PB_{S_k} = E\left(\sum_{i=0}^{k} \rho_i + \sum_{j=k+1}^{n-1} \sum_{i=0}^{k} \frac{\rho_i}{\sum_{p=0}^{k} \rho_p} \rho_k'(\delta_i - \delta_k), W\right). \qquad (3.96)$$

An iterative method can be used to solve all the coupled equations to obtain the BLP of each class considering the interference among the classes [8].

3.9.1 Generic Model Without Class Isolation

As mentioned earlier, although the assumption of class isolation and conservation law in multi-class OBS networks facilitates the use of classical Markov chain models for blocking probability, the practicality and accuracy of these models is questionable in many situations. This motivates a more general model for blocking probability in multi-class OBS networks that is accurate for any degree of isolation and systems that do not obey the conservation law. Supporting this motivation is the fact that there are several burst assembly mechanisms and scheduling algorithms which change the burst length and offset time distribution to a large extent. Therefore, the model must consider the offset time and burst length distribution explicitly. Unlike the recursive model presented above, a generic model for blocking of OBS node supporting multiple classes of traffic and with bursts of any size distribution is presented next. The analysis works well for any degree of isolation between the traffic classes and for either work-conserving or non-work-conserving systems [3, 4].

Assume that there are n classes of traffic such that class j has a higher priority over i if $1 \le i < j \le n$. The main aim of the analysis is to find the average blocking probability of a class j burst, denoted by PB_j. In JET protocol, a BHP arriving ahead of the data burst attempts to reserve bandwidth for a fixed duration equal to the length of the burst. Let the time slot requested be denoted by T_s. When the BHP finds one or more bursts already using either a portion or entire T_s, the data burst is blocked. Let s_j^k indicate that k bursts have reserved the bandwidth in T_s of a class j burst. Then the blocking probability of class j burst can be written as

$$PB_j = 1 - P[\text{no bursts are reserved during } T_s \text{ of class } j \text{ burst}]$$

$$= 1 - \left(P[s_j^0] + \sum_{k=1}^{\infty} P[s_j^k \cap \text{all } k \text{ bursts are blocked}] \right)$$

$$= 1 - P[s_j^0] - \sum_{k=1}^{\infty} \left(P[k \text{ bursts blocked} | s_j^k] . P[s_j^k] \right). \qquad (3.97)$$

To evaluate the above equation, consider the blocking process of two bursts. A class j burst is blocked by class i burst whenever one of the following events occurs:

- The BHP of class i burst arrives before the BHP of class j burst.
- The class i burst is still under transmission when the class j burst arrives for transmission; this event corresponds to head segmentation.
- The class i burst arrives before the end of class j burst; this event where i has higher priority over j corresponds to the tail segmentation.

Let E_{ji} denote the intersection of all the three events mentioned above where a class j burst is blocked, whose probability can be written as

$$P[E_{ji}] = P[(a_i < a_j) \cap (a_i + o_i + L_i < a_j + o_j) \cap (a_i + o_i < a_j + o_j + L_j)], \quad (3.98)$$

where a_i, o_i, and L_i are the arrival time, offset time, and the mean length of bursts of class i, respectively.

Assuming that the bursts of class i arrive following a Poisson distribution, by the memoryless property [20],

$$P[E_{ji}] = P[(a_i - a_j) \in \{\delta_{ji}\} - L_i, \min\{0, L_j + \delta_{ji}\}\}]$$
$$= P[\alpha_i < \min\{0, L_j + \delta_{ji}\} - (\delta_{ji}] - L_i)], \qquad (3.99)$$

where $\delta_{ji} = o_j - o_i$ is the difference between the offset times of bursts of class j and class i, and α_i is a random variable to represent the inter-arrival time of class i bursts. It is easy to write the probability of the complement of E_{ji} using the above equation as

$$P[\overline{E_{ji}}] = P[\alpha_i > \min\{0, L_j + \delta_{ji}\} - (\delta_{ji}) - L_i)]$$

$$= \begin{cases} P[\alpha_i > L_i + L_j], & \delta_{ji} < 0, L_j < -\delta_{ji} \\ P[\alpha_i > L_i - \delta_{ji}\}], & \delta_{ji} < 0, L_j \ge -\delta_{ji} \\ P[\alpha_i > L_i - \delta_{ji}\}], & \delta_{ji} \ge 0. \end{cases} \qquad (3.100)$$

Combining the first two cases in the above equation by conditioning over all possible values of L_j gives

$$P[\overline{E_{ji}}] = \begin{cases} \int_0^{-\delta_{ji}} P[\alpha_i > L_i + l_j] f_{L_j}(l_j) dl_j \\ + \int_{-\delta_{ji}}^{\infty} P[\alpha_i > L_i - \delta_{ji}\}] f_{L_j}(l_j) dl_j, & \delta_{ji} < 0 \\ P[\alpha_i > L_i - \delta_{ji}\}], & \delta_{ji} \ge 0 \end{cases}$$

$$= \begin{cases} \int_0^{-\delta_{ji}} P[\alpha_i > L_i + l_j] f_{L_j}(l_j) dl_j \\ + P[\alpha_i > L_i - \delta_{ji}\}](1 - F_{L_j}(-\delta_{ji})), & \delta_{ji} < 0 \\ P[\alpha_i > L_i - \delta_{ji}\}], & \delta_{ji} \ge 0, \end{cases} \qquad (3.101)$$

where $f_{L_j}(l)$ and $F_{L_j}(l)$ are the probability density and distribution functions for the burst length of class j, respectively. To evaluate $P[\alpha_i > L_i + l_j]$ and $P[\alpha_i > L_i - \delta_{ji}\}]$ the following identity for two non-negative random variables X and Y with joint distribution function $f_{XY}(x, y)$ can be used [20]:

$$P[X > Y + c] = \begin{cases} \int_0^{\infty} \int_0^{x-c} f_{XY}(x, y) dy\, dx, & c < 0 \\ \int_0^{\infty} \int_{y+c}^{\infty} f_{XY}(x, y) dx\, dy, & c \ge 0, \end{cases} \qquad (3.102)$$

where c is a constant.

This completes the analysis of the blocking process with two bursts contending in a time slot of T_s. Next, continuing with the evaluation of general multi-class blocking, note that in Eq. 3.97, $P[k \text{ bursts blocked}|s_j^k] \ll P[s_j^0]$ because blocking events are usually fewer. Therefore, neglecting all the higher order terms, Eq. 3.97 can be simply written as

$$\text{PB}_j \approx 1 - P[s_j^0] = 1 - P\left[\bigcap_{i=1}^{n} \overline{E_{ji}} \right]$$

$$= 1 - P\left[\bigcap_{i=1}^{j-1} \overline{E_{ji}} \bigcap_{i=j}^{n} \overline{E_{ji}} \right]. \qquad (3.103)$$

In obtaining the second step of the above equation, the intersection term is separated into the terms corresponding to the effects of higher priority traffic and the effects of equal and low-priority traffic. From Eq. 3.101, it can be observed that when $\delta_{ji} \ge 0$, since $\overline{E_{ji}}$ depends only on α_i and L_i, all the events are independent. Assuming that the $\overline{E_{ji}}$ terms are independent for even $\delta_{ji} < 0$,

$$PB_j = 1 - \prod_{i=1}^{j-1} P[\overline{E}_{ji}] \prod_{i=j}^{n} P[\overline{E}_{ji}]. \qquad (3.104)$$

Substituting Eq. 3.101 into Eq. 3.104 gives

$$PB_j = 1 - \prod_{i=1}^{j-1} (\int_0^{-\delta_{ji}} P[\alpha_i > L_i + l_j] f_{L_j}(l_j) dl_j$$

$$+ [1 - F_{L_j}(-\delta_{ji}) P[\alpha_i > L_i - \delta_{ji}]]) \cdot \prod_{i=j}^{n} P[\alpha_i > L_i - \delta_{ji}]. \qquad (3.105)$$

The above equation gives the blocking probability of a class j burst when there are n classes of traffic in the network for any general burst length distribution and offset time. There is no assumption of isolation among different classes of traffic and the network need not be work-conserving. To demonstrate the advantage of such a model, consider that the bursts have exponentially distributed length. Then the probability density function and the distribution function for class j bursts are given by

$$f_{L_j}(l) = \frac{1}{\overline{L}_j} e^{-l/\overline{L}_j} \text{ and } F_{L_j}(l) = 1 - e^{-l/\overline{L}_j}, \qquad (3.106)$$

where $\overline{L}_j$ is the mean length of bursts belonging to class j. Substituting Eq. 3.106 into Eq. 3.105 and using the identity given below for two independent exponential variables X and Y with corresponding mean values $1/a_x$ and $1/a_y$,

$$P[X > Y + c] = \begin{cases} \frac{a_y + a_x(1 - e^{a_y c})}{a_y + a_x} & c < 0 \\ \frac{a_y e^{-a_x c}}{a_y + a_x} & c \geq 0 \end{cases} \qquad (3.107)$$

the average blocking probability of bursts belonging to class j with exponentially distributed length can be obtained as

$$PB_j = 1 - \frac{\prod_{i-1}^{j-1} \left[\frac{1 + \lambda_i \overline{L}_j e^{\delta_{ji}(\lambda_i + 1/\overline{L}_i)}}{1 + \overline{L}_j \lambda_i} \right] \prod_{i=j}^{n} \left[1 + \rho_i (1 - e^{-\delta_{ji}/\overline{L}_i}) \right]}{\prod_{i=1}^{n} [1 + \rho_i]}, \qquad (3.108)$$

where λ_j is the mean arrival rate of bursts of class j and $\rho_j = \lambda_j \overline{L}_j$ is the offered load due to bursts of class j. Note that primarily the blocking probability of bursts is a function of the term $\delta_{ij}/\overline{L}_j$, the degree of isolation between two classes i and j, which is also observed from the earlier analysis. Simulations in [3] show that the analysis is accurate for any degree of isolation and for a large range of load values. As the model does not require class isolation, it holds good for any length and distribution of bursts which is otherwise not possible with models that assume class isolation. However, in general most of the analysis presented in this section which does not assume class isolation is not accurate for large blocking or high load values although the error is small. But since most of the networks operate with blocking lesser than 10%, this error can be neglected.

References

1. Akar, N., Karasan, E.: Exact calculation of blocking probabilities for bufferless optical burst switched links with partial wavelength conversion. In: Proceedings of BROADNETS, pp. 110–117 (2004)
2. Akar, N., Karasan, E., Dogan, K.: Wavelength converter sharing in asynchronous optical packet/burst switching: An exact blocking analysis for Markovian arrivals. IEEE Journal on Selected Areas in Communications **24**(12), 69–80 (2006)
3. Barakat, N., Sargent, E.H.: Performance analysis of optical burst switching networks with and without class isolation. In: Proceedings of IEEE GLOBECOM, pp. 2513–2518 (2003)
4. Barakat, N., Sargent, E.H.: An accurate model for evaluating blocking probabilities in multi-class optical burst switching systems. IEEE Communications Letters **8**(2), 119–121 (2004)
5. Barry, R.A., Humblet, P.A.: Models of blocking probability in all-optical networks with and without wavelength changers. IEEE Journal on Selected Areas in Communications **14**(5), 858–867 (1996)
6. Birman, A.: Computing approximate blocking probabilities for a class of all-optical networks. IEEE Journal on Selected Areas in Communications **14**(5), 852–857 (1996)
7. David, H., Nagaraja, H.: Order Statistics. Wiley, USA (2003)
8. Dolzer, K., Gauger, C.: On burst assembly in optical burst switching networks – A performance evaluation of just-enough-time. In: Proceedings of International Teletraffic Congress, pp. 1–12 (2001)
9. Dolzer, K., Gauger, C., Spath, J., Bodamer, S.: Evaluation of reservation mechanisms for optical burst switching. AEU Journal of Electronics and Communications **55**(1) (2001)
10. Durhuus, T., Mikkelsen, B., Joergensen, C., Danielsen, S.L., Stubjaer, K.E.: All-optical wavelength conversion by semiconductor optical amplifiers. Journal of Lightwave Technology **14**(6), 942–954 (1996)
11. Eramo, V., Listanti, M., Pacifici, P.: A comparison study on the wavelength converters number needed in synchronous and asynchronous all-optical switching architectures. Journal of Lightwave Technology **21**(2), 340–355 (2003)
12. Gauger, C.M.: Performance of converter pools for contention resolution in optical burst switching. In: Proceedings of SPIE Opticomm, pp. 109–117 (2002)
13. Golub, G.H., van Loan, C.F.: Matrix Computations. The John Hopkins University Press, USA (1996)
14. Gross, D., Harris, C.: Fundamentals of Queueing Theory. Wiley-Interscience, USA (1997)
15. Hernandez, J.A., Aracil, J., Pedro, L., Reviriego, P.: Analysis of blocking probability of data bursts with continuous-time variable offsets in single-wavelength OBS switches. Journal of Lightwave Technology **26**(12), 1559–1568 (2008)
16. Kaheel, A., Alnuweiri, H., Gebali, F.: Analytical evaluation of blocking probability in optical burst switching networks. In: Proceedings of IEEE GLOBECOM, pp. 1548–1553 (2004)
17. Kaheel, A., Alnuweiri, H., Gebali, F.: A new analytical model for computing blocking probability in optical burst switching networks. In: Proceedings of 9th ISCC, pp. 264–269 (2004)
18. Kaheel, A., Alnuweiri, H., Gebali, F.: A new analytical approach to compute blocking probability in optical burst switching networks. IEEE Journal on Selected Areas in Communications **24**(12), 120–128 (2006)

19. Kelly, F.P.: Blocking probabilities in large circuit switched networks. Advances in Applied Probability **18**, 473–505 (1986)
20. Kleinrock, L.: Queueing Systems. Vol. 1, John Wiley and Sons, USA (1975)
21. Kovacevic, M., Acampora, A.: Benefits of wavelength translation in all-optical clear channel networks. IEEE Journal on Selected Areas in Communications **14**(5), 868–880 (1996)
22. Lee, K., Li, V.: A wavelength convertible optical network. Journal of Lightwave Technology **11**(5), 962–970 (1993)
23. Mingwu, Y., Aijun, W., Zengji, L.: Blocking probability of asynchronous optical burst/packet switches with limited range wavelength conversion. IEEE Photonics Technology Letters **18**(12), 1302–1304 (2006)
24. Mingwu, Y., Zengji, L., Aijun, W.: Accurate and approximate evaluations of asynchronous tunable wavelength converter sharing schemes in optical burst switched networks. Journal of Lightwave Technology **23**(10), 2807–2815 (2005)
25. Munkres, J.R.: Elements of Algebraic Topology. Persues Press, USA (1993)
26. Puttasubbappa, V.S., Perros, H.G.: Performance analysis of limited-range wavelength conversion in an OBS switch. Telecommunication Systems **31**, 227–246 (2006)
27. Rosberg, Z., Vu, H.L., Zukerman, M., White, J.: Blocking probabilities of optical burst switching networks based on reduced fixed point approximations. In: Proceedings of IEEE INFOCOM, pp. 2008–2018 (2003)
28. Rosberg, Z., Vu, H.L., Zukerman, M., White, J.: Performance analyses of optical burst-switching networks. IEEE Journal on Selected Areas in Communications **21**(7), 1187–1196 (2003)
29. Rosberg, Z., Zalesky, A., Vu, H.L., Zukerman, M.: Analysis of OBS networks with limited wavelength conversion. IEEE Transactions on Networking **14**(5), 1118–1127 (2006)
30. Ross, S.M.: Introduction to Probability Models. Academic Press, USA (2006)
31. Shalaby, H.M.H.: A simplified performance analysis of optical burst-switched networks. Journal of Lightwave Technology **25**(4), 986–995 (2007)
32. Sharma, V., Varvarigos, E.A.: An analysis of limited wavelength translation in regular all-optical WDM network. Journal of Lightwave Technology **18**(12), 1606–1619 (2000)
33. Stewart, W.J.: Introduction to the Numerical Solution of Markov Chains. Princeton University Press, USA (1994)
34. Tripathi, T., Sivarajan, K.N.: Computing approximate blocking probabilities in wavelength routed all-optical networks with limited range wavelength conversion. IEEE Journal on Selected Areas in Communications **18**(10), 2123–2129 (2000)
35. Vu, H.L., Zukerman, M.: Blocking probability for priority classes in optical burst switching networks. IEEE Communications Letters **6**(5), 214–216 (2002)
36. Yates, J., Lacey, J., Everitt, D., Summerfield, M.: Limited range wavelength translation in all-optical networks. In: Proceedings of IEEE INFOCOM, pp. 954–961 (1996)
37. Yoo, M., Qiao, C.: Supporting multiple classes of service in IP over WDM networks. In: Proceedings of IEEE GLOBECOM, pp. 1023–1027 (1999)
38. Yoo, M., Qiao, C., Dixit, S.: QoS performance of optical burst switching in IP-over-WDM networks. IEEE Journal on Selected Areas in Communications **18**(10), 2062–2071 (2000)
39. Zalesky, A., Vu, H.L., Zukerman, M., Rosberg, Z., Wong, E.W.M.: Evaluation of limited wavelength conversion and deflection routing as methods to reduce blocking probability in optical burst switched networks. In: Proceedings of IEEE ICC, pp. 1543–1547 (2004)
40. Zeng, G., Lu, K., Chlamtac, I.: On the conservation law in optical burst switching networks. In: Proceedings of SPECTS, pp. 124–129 (2004)

Chapter 4
Contention Resolution in OBS Networks

4.1 Introduction

OBS networks using the JET/JIT reservation mechanism, which is a one-way and unacknowledged reservation mechanism, provide connectionless switching. Due to this feature, the bursts may contend with one another at the core nodes. Contention among two bursts occurs due to the overlap of two or more bursts (in time) that arrive simultaneously on two different links or wavelengths and request the same wavelength at a given time. In electronic packet switching networks, contention is handled by buffering. However, in optical networks, buffers are difficult to implement and there is no optical equivalent of random access memory. When multiple bursts contend for a wavelength at an intermediate node, all but one of the contending bursts is dropped. Burst loss due to contention is a major source of concern in OBS networks. Since these losses are quite random in nature and occur due to lack of buffers, it poses a serious problem in the deployment of OBS technology in networks that support QoS at the higher layers.

To resolve contention losses, several mechanisms have been proposed among which the main ones are wavelength conversion, burst segmentation, deflection routing, and optical buffering [45]. The effect of wavelength conversion on the blocking probability at a node was discussed in the previous chapter. In this chapter, the impact of burst segmentation, deflection routing, and optical buffering using FDLs on the blocking probability is analyzed and their role in reducing contention losses is studied. Different theoretical models are discussed which help to analyze the benefit of these methods in reducing the blocking probability at the core nodes.

4.2 Burst Segmentation

When two or more bursts contend for the same wavelength at the same time, all of them except the one, which arrives first or which has the highest priority, are usually dropped. Instead of dropping the entire burst, if the burst that arrives later is switched onto the wavelength after the transmission of the current burst at least a part of the burst can be sent. Since a burst contains many IP packets, at least a few of the packets can be delivered at the destination with such an approach. This

T. Venkatesh, C. Siva Ram Murthy, *An Analytical Approach to Optical Burst Switched Networks*, DOI 10.1007/978-1-4419-1510-8_4,
© Springer Science+Business Media, LLC 2010

approach which is known as burst segmentation (BS) improves the packet delivery rate at the IP layer [46]. BS has a few advantages over dropping the entire burst. It reduces the delay due to retransmission and burst assembly by delivering at least some of the packets. If the two bursts overlap only for a short duration, BS helps the delivery of almost the entire burst. It reduces the packet loss probability which is not same as BLP if a portion of the burst is delivered. Certain applications that have stringent delay and loss requirements are greatly benefited by the BS. Another proposal similar to the BS is known as optical composite burst switching (OCBS) [14]. Similar to the BS, a switch with the OCBS scheme discards only the initial part of the burst till a wavelength on the outgoing fiber is available; from that instant the remaining burst is transmitted. The difference between OCBS and BS is that the latter scheme has two flavors: head and tail segmentation. In head segmentation, the head of the burst that arrives later is segmented. In tail segmentation, the tail of the burst in transmission is segmented whereas the burst that arrives later is transmitted completely. If the tail is segmented, the possibility of IP packets being sent without the headers is minimized and the out-of-order delivery of packets (if retransmitted by the higher layer) is also reduced [46]. However, in tail segmentation the BHP contains the original length of the burst which wastes some bandwidth. To overcome this problem, the core node where a burst is segmented has to send a trailer packet to the other nodes along the path to the destination which updates the burst length information. On the other hand, head segmentation has the advantage that once a burst reaches a node without contention, it is guaranteed to be transmitted.

4.2.1 Blocking Analysis of a Node with Burst Segmentation

The blocking probability of a node that allows BS or OCBS cannot be computed using the Erlang B formula given earlier in Eq. 3.1. At a node without OCBS, a burst is either accepted or rejected in its entirety so that the packet loss probability is equal to the BLP. To model the blocking at a node with OCBS, an $M/G/\infty$ queue with an unlimited number of pseudo-servers and w real servers is used [35]. With this model that uses the $M/G/\infty$ queue, whenever the system is full (i.e., all the w wavelengths are transmitting bursts) every new burst that arrives is accepted by the $w+1$th server which is a pseudo-server. For each subsequent burst that arrives when the system is full, the number of active pseudo-servers increases by 1 as shown in Fig. 4.1. The pseudo-servers serve the bursts till one of the real servers is free. As long as the bursts are being served by the pseudo-servers, they are being segmented. When one of the real servers becomes free, the first burst being served by a pseudo-server (the first segmented burst) is moved to the real server so that the remaining portion of the burst is transmitted.

As shown in Fig. 4.1, the shaded portion of the burst in pseudo-servers is dropped due to its overlap with the other bursts being served by one of the w real servers. Whenever one of the real servers is free, the remaining portion of the truncated

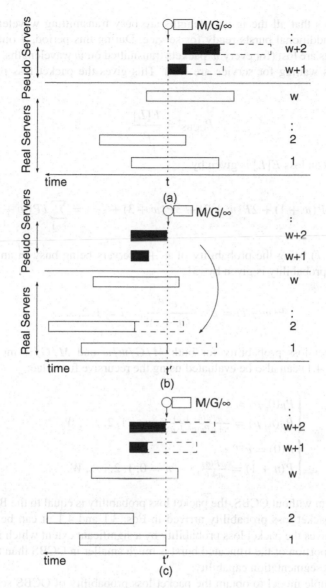

Fig. 4.1 Pseudo-server model for an $M/G/\infty$ queue

burst is allocated to this free server. The OCBS system is completely analogous to the system where the pseudo-server becomes a real server as soon as one of the real servers becomes free. Using $M/G/\infty$ model for OCBS gives a simple way to calculate the packet loss probability. The case when the number of active servers is less than w is similar to the case of w or fewer wavelengths transmitting bursts. This represents the case of no burst loss. The case of $w + j, j > 1$ servers being busy indicates that all the w wavelengths are busy transmitting wavelengths and there are j additional bursts ready for service. During this period, j out of every $w + j$ packets are lost (for every w packets transmitted on w wavelengths, j packets of the bursts waiting for service are lost). This gives the packet loss probability for OCBS as

$$P_{\text{OCBS}} = \frac{E[L]}{\rho}, \tag{4.1}$$

where the mean loss $E[L]$ is given by

$$E[L] = 1P(w + 1) + 2P(w + 2) + 3P(w + 3) + \cdots = \sum_{i=1}^{\infty} i P(w + i), \tag{4.2}$$

with $P(w + i)$ being the probability of $w + i$ servers being busy in an $M/G/\infty$ queue. This probability is given by [3]

$$P(w + i) = \rho^{w+i} \frac{e^{-\rho}}{(w + i)!}, \quad i = 1, 2, \ldots. \tag{4.3}$$

The packet loss probability for both $M/G/w/w$ and $M/G/\infty$ models (i.e., Eqs. 3.1 and 4.1) can also be evaluated using the recursive formulae:

$$\begin{cases} P_B(0, \rho) = 1 \\ P_B(n, \rho) = \frac{\rho P_B(n-1,\rho)}{n+\rho P_B(n-1,\rho)}, & n = 1, 2, \ldots, W, \end{cases} \tag{4.4}$$

$$\begin{cases} P(0) = e^{-\rho} \\ P(n + 1) = \frac{\rho P(n)}{n+1}, & n = 0, 1, 2, \ldots, W, \ldots. \end{cases} \tag{4.5}$$

In a system without OCBS, the packet loss probability is equal to the BLP. Comparing the packet loss probability arrived in Eqs. 3.1 and 4.1, it can be seen that OCBS improves the packet loss probability by a significant extent which is because the average portion of the truncated burst is much smaller in OCBS than that in the case without segmentation capability.

In the simple model to obtain the packet loss probability of OCBS scheme discussed above, it is assumed that the bursts are always segmented in integral units of packets. That is, when a burst is segmented no headers of the packets are lost and the portion of the burst received at the egress node is completely useful. If a packet is segmented in the middle, the useless portion of the packet (the portion without a header) is also assumed to be lost. This model neglects the loss of packets during the switching time and those packets lost due to the transmission of partial bursts. These assumptions implicate that the IP packets are smaller in size compared to the size of the bursts. This assumption is fair to a certain extent if we assume that each burst contains several hundreds of packets.

4.2.1.1 Analysis for Small Burst Sizes

To evaluate the IP packet loss probability exactly when the packet size is not negligible relative to the burst size, the length of the burst received which is useful as well as useless (the IP packets that have been partially segmented) needs to be considered [14]. To avoid dealing with the continuous distribution of the burst length, assume that the traffic is from a collection of the ON/OFF sources where the ON period corresponds to the length of the burst L_b. Let the OFF period be represented by O. Let the nth ON and OFF periods be iid, denoted by L_b^n and O^n, respectively. Let their average values be denoted by L_b and O, respectively, so that the traffic offered on a wavelength is given by $\rho = L_b/(L_b + O)$. With burst segmentation at the core node, let $L_{b,a}^n$ be the length of the nth burst that is received which would be L_b^n if the entire burst is received. Now, $L_{b,a}^n$ consists of two portions, one with a length $L_{b,a,u}^n$ which is the useful part in which all the IP packets are complete and the other one with a length $L_{b,a,nu}^n$ which is the useless part with partial IP packets. The useless part of the burst would be discarded at the destination node but it contributes to the load along the path. Since the random variable L_b^n, $n = 1, 2, \ldots$ is iid, with an expected value of $\overline{L_b}$, it follows that $L_{b,a,u}^n$, $n = 1, 2, \ldots$ and $L_{b,a,nu}^n$, $n = 1, 2, \ldots$ are also iid, with the expected values $\overline{L_{b,a,u}}$ and $\overline{L_{b,a,nu}}$, respectively.

Let ρ_a be the fraction of the offered load carried (accepted) by the outgoing link and ρ_r the fraction of the offered load rejected by the link which is related to ρ_a by $\rho_r = 1 - \rho_a$. ρ_a can be evaluated using the processes $\{l_b^n, n = 1, 2, \ldots\}$ and $\{l_{b,a}^n, n = 1, 2, \ldots\}$ (corresponding to the random variables L_b^n, $n = 1, 2, \ldots$ and $L_{b,a}^n$, $n = 1, 2, \ldots$) as

$$\rho_a = \lim_{k \to \infty} \frac{\sum_{n=1}^{k} l_{b,a}^n}{\sum_{n=1}^{k} l_b^n}. \tag{4.6}$$

By expressing the length of the accepted burst as the sum of useful and useless parts and taking $L_{b,a,nu}^n$ to be zero when either a burst is rejected (i.e., $L_{b,a}^n = 0$) or a burst is entirely accepted (i.e., $L_b^n = L_{b,a}^n$), ρ_a can be written as

$$\rho_a = \lim_{k \to \infty} \frac{\sum_{n=1}^{k} l_{b,a,nu}^n \delta(l_b^n - l_{b,a}^n)\, \delta(l_{b,a}^n) + \sum_{n=1}^{k} l_{b,a,u}^n}{\sum_{n=1}^{k} l_b^n} \tag{4.7}$$

$$= \lim_{k \to \infty} \frac{\sum_{n=1}^{k} l_{b,a,u}^n}{\sum_{n=1}^{k} l_b^n} + \lim_{k \to \infty} \frac{\sum_{n=1}^{k} l_{b,a,nu}^n \delta(l_b^n - l_{b,a}^n)\, \delta(l_{b,a}^n)}{\sum_{n=1}^{k} l_b^n}.$$

Therefore, ρ_r can be written as

$$\rho_r = 1 - \rho_a = P_p - \lim_{k \to \infty} \frac{\sum_{n=1}^{k} l_{b,a,nu}^n \delta(l_b^n - l_{b,a}^n)\, \delta(l_{b,a}^n)}{\sum_{n=1}^{k} l_b^n}, \tag{4.8}$$

where P_p is the packet loss probability which is the ratio of the length of the useful part of the burst received to the total burst length. If P_d is the probability that an arriving burst is dropped, then the second term of Eq. 4.8 can be written as $P_d \frac{\overline{L_{b,a,nu}}}{L_b}$. P_d can be written as $P_d = 1 - (P_a + P_r)$, where P_a is the probability that a burst is entirely accepted and P_r is the probability that a burst is rejected.

A new burst is accepted completely if it finds less than w active input wavelengths carrying bursts. Therefore, P_a can be written as

$$P_a = \sum_{i=1}^{w} \binom{w}{i} \rho^i (1 - \rho)^{w-i}. \tag{4.9}$$

The probability that a burst is rejected can be obtained by using the law of total probability as follows:

$$P_r = \sum_{i=w}^{N.w} P_{r|i} P_i, \tag{4.10}$$

where N is the number of input fibers, $P_{r|i}$ is the probability that a burst is rejected provided that an arriving burst finds i bursts already in transmission, i.e., $P_i = \binom{N.w}{i} \rho^i (1 - \rho)^{N.w-i}$. $P_{r|i}$ can simply be written as

$$P_{r|i} = \begin{cases} 0, & \text{if } i < w \\ \sum_{j=w}^{i} \binom{i}{j} P_c^j (1 - P_c)^{i-j}, & \text{if } i \geq w \end{cases}, \tag{4.11}$$

where P_c is the probability that the residual length of the burst in transmission is greater than the length of the arriving burst which can be written as

$$P_c = \int_0^\infty P_{c|x} f_{L_b}(x) dx, \tag{4.12}$$

where $P_{c|x}$ is conditioned on the length of the burst arriving in the interval $[x, x+dx]$ given by [14]

$$P_{c|x} = \frac{1}{\overline{L_b}} \left(1 - \int_0^x f_{L_b}(y) dy \right). \tag{4.13}$$

Using all the expressions developed above, the packet loss probability in the OCBS scheme is approximately given by

$$P_p = \rho_r + P_d \frac{\overline{L_{b,a,nu}}}{\overline{L_b}}. \tag{4.14}$$

4.2.2 Benefit of Burst Segmentation

As discussed earlier, burst segmentation (or OCBS) improves the throughput in the network in terms of the number of IP packets successfully delivered by recovering at least a portion of the burst. In fact under low load conditions (or when the number of wavelengths is high) and with wavelength conversion, the improvement due to the burst segmentation is high. Simple numerical evaluation of the blocking probability with and without segmentation (using Eq. 4.4) shows that even without wavelength conversion the benefit due to segmentation is significant. This is mainly due to the fact that on an average, the truncated portion of the burst is significantly smaller than the portion successfully transmitted. Under the light load conditions, a simple expression for the relative benefit of BS over JET is given here. Recall that the packet loss probability using JET is given by Eq. 3.1 while for OCBS (or BS) it is given by Eq. 4.1. The relative benefit of OCBS over JET is given by the ratio

$$\frac{P_B(\rho, w)}{P_{OCBS}(\rho, w)} = \frac{P_B(\rho, w)}{E[L]/\rho}. \tag{4.15}$$

Observe from Eq. 4.3 that $P(w+i)$ can be written as $[\rho/(w+i)].P(w+i-1), i \geq 1$. When the load is low or $w \gg \rho$, we can write $P(w+1) \gg P(w+2) \gg P(w+3) \gg \ldots$, so that $E[L] \approx P(w+1)$. Hence, the relative benefit of BS over JET can be written as

$$\frac{P_B(\rho, w)}{P_{OCBS}(\rho, w)} = \frac{P_B(\rho, w)}{P(w+1)/\rho} = \frac{w+1}{e^{-\rho} \sum_{i=0}^{w} \frac{\rho^i}{i!}} \tag{4.16}$$

and

$$\lim_{w \to \infty} \frac{P_B(\rho, w)}{P(w+1)/\rho} = w + 1. \tag{4.17}$$

This result although applicable only with an unrealistically low load shows that there is a definite benefit with burst segmentation and it increases with the number of wavelengths. For the case of $\rho/w < 1$, the asymptotic value of the ratio in Eq. 4.17 is approximately given by $O(w)$ with a constant $w \frac{(1-(\frac{\rho}{w})^2)}{\rho}$ [41]. That means BS offers a factor of $w^2 \frac{(1-(\frac{\rho}{w})^2)}{\rho}$ improvement over JET in terms of blocking probability which can be around two orders of magnitude for dense WDM (DWDM) systems with hundreds of wavelengths.

To gain deeper and more quantitative understanding of the benefit of burst segmentation under realistic load conditions, a probabilistic model is presented for burst segmentation in this section which compares the statistical characteristics of the truncated part of the burst with the part transmitted successfully [36].

4.2.2.1 Probabilistic Model for Burst Segmentation

Assume that bursts arrive following a homogeneous Poisson process with rate λ and the burst duration is exponentially distributed with mean $1/\mu$. Assume that only one burst arrives during τ which is the duration between the epochs when the system (a link) is busy and the first subsequent departure of a burst. The assumption that there is only one burst arrival when the system is full is valid for low loss probabilities. Such a burst is lost if none of the wavelengths becomes free for the entire duration of the burst else it is truncated.

Let the random variable S denote the time of overlap between τ and the service time corresponding to the truncated portion of the burst and R be the random variable indicating the time of overlap between τ and the portion of the burst successfully transmitted. $S + R$ is essentially the random variable indicating the total duration of the burst and R is the *residual service time*. The variables are illustrated in Fig. 4.2 where as explained earlier, only one burst arrives during the time interval τ in which all the w wavelengths are busy. The duration for which the burst is segmented is indicated by S and the residual service time, during which the burst is transmitted, is shown as R.

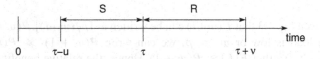

Fig. 4.2 Illustration of probabilistic model for burst segmentation

Since we assume homogeneous Poisson arrival of bursts, the elementary conditional probability that the first burst arrives in the interval $[\tau - u, \tau - u + du]$, $0 < u < \tau$, is given by $e^{-\lambda(\tau-u)}\lambda du$ and the elementary conditional probability that the service time of the burst exceeds u and it departs in the interval $[\tau+v, \tau+v+dv]$ is

$$P = e^{-\lambda(\tau-u)}\lambda du \; e^{-\mu(u+v)}\mu dv, \tag{4.18}$$

where $[\tau, \tau + d\tau]$ is the interval when all the w wavelengths are busy. Note that τ is an exponentially distributed random variable with mean $w\mu$ so that the elementary probability that there is an overlapping burst such that $u < S < u + du$ and $v < R < v + dv$ is given by

$$P\{O, S \in (u + du), R \in (v + dv)\} = \int_u^\infty Pe^{-w\mu\tau}w\mu \; d\tau \tag{4.19}$$

$$= \frac{w\rho}{(w + 1)(w + \rho)} \; e^{-(w+1)\mu u}(w + 1)\mu du \; e^{-\mu v}\mu dv,$$

where P is defined in Eq. 4.18 and $\rho = \lambda/\mu$. The first factor $P\{O\} = \frac{w\rho}{(w+1)(w+\rho)}$ is the probability that there is an overlapping burst during τ.

Equation 4.19 proves that the overlap time S and the residual service time R are independent and are exponentially distributed with parameters of $(w + 1)\mu$ and μ,

respectively. It can also be seen that the mean values of the lost portion of the burst
(S) and the portion successfully transmitted (R) are related by

$$\frac{E[S]}{E[R]} = \frac{1}{w+1}.\tag{4.20}$$

This result is similar to Eq. 4.17 but gives better insight into the relation between
the truncated part and the residual service time of burst being segmented. When the
system is full, the portion of the burst lost in segmentation on an average is always
less than that successfully transmitted. Further, as w increases, the ratio $E[S]/E[R]$
decreases which indicates that fewer packets are lost in segmentation so that packet
delivery is better.

4.2.3 Prioritized Burst Segmentation

In the burst segmentation scheme, either the burst being serviced is segmented when
a new burst contends for the same duration or the contending burst is segmented till
the wavelength is free. To minimize the lost portion of the burst, the length of the
contending burst is compared with the residual length of the burst under service and
the one with the smaller length is dropped. A simple extension of this scheme which
includes bursts with priorities is discussed in this section. In this scheme, a priority
field included in the burst header is used to decide which burst gets segmented.
Priority-based burst segmentation or prioritized segmentation [44] can be used to
provide service differentiation in terms of packet loss probability in applications
which require loss guarantees. Unlike the method which uses an extra offset time
to provide lower loss probability to some bursts [50], prioritized segmentation helps
to provide loss differentiation among different classes of the traffic without extra
delay. In addition to prioritized segmentation, assembling the packets from different
classes in different portions of the burst, termed as composite burst assembly [44],
helps to meet the loss guarantees for multiple classes of traffic.

Figure 4.3 illustrates the four cases of contention that may arise among bursts
with two levels of priority. The burst that is currently being serviced is termed as
the *original burst* while the burst that arrives later is termed as the *contending burst*
denoted by B_o and B_c, respectively, with corresponding priorities denoted by P_o
and P_c. If the contending burst has a higher priority, it wins the contention and
the original burst is segmented as shown in case (i) (this is the typical case of tail
segmentation) while case (ii) shows the original burst with a higher priority winning
the contention causing the contending burst to be segmented (head segmentation).
When the two bursts have equal priority, then the segmentation is done by comparing the length of the contending burst with the residual length of the original
burst. If the length of the contending burst is larger, the original burst is segmented
(shown in case (iii) of Fig. 4.3), whereas if the contending burst is smaller, it is
dropped.

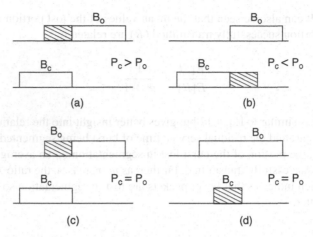

Fig. 4.3 Illustration of prioritized burst segmentation

This section evaluates the packet loss probability for both the high- and low-priority bursts. A higher priority burst segments the lower priority burst in service which is similar to preempting the burst in service. If the burst in service has a higher priority, then the contending burst is dropped. Assume that all the bursts have equal offset time so that there is no retro-blocking. Also assume that for each source–destination pair sd, bursts with high and low priority arrive according to a Poisson process with a mean arrival rate of λ_{sd}^h and λ_{sd}^l, respectively. Further, assume that the switching time is very small and loss due to partial packets received at the egress is neglected.

The method followed in this section to compute the packet loss probability due to segmentation is to compare the distribution of the burst length at the egress node with that at the ingress node. Let $F_{0,sd}^h$ denote the cumulative distribution function (cdf) of the initial burst length at the ingress node (denoted by 0th hop) for a source–destination pair. At the kth hop the cdf of the burst length is denoted by $F_{k,sd}^h$. Let $H_{k,sd}^h$ be the cdf of the arrival time of the next high-priority burst on the link at kth hop between the source–destination pair sd which can be written as

$$H_{k,sd}^h(t) = 1 - e^{-\lambda_{k,sd}^h t}, \tag{4.21}$$

where $\lambda_{k,sd}^h$ is the arrival rate of all the high-priority bursts at the kth hop between the source–destination pair sd. Since the original burst is segmented only on the arrival of a burst with higher priority, the probability that the burst length is less than or equal to t after the first hop equals to the probability of the burst being already of length lesser than or equal to t or the probability of a higher priority burst arriving in time less than or equal to t. This gives the equation for computation of cdf of the burst length at the first hop as

$$F_{1,sd}^h(t) = 1 - \left(1 - F_{0,sd}^h(t)\right)\left(1 - H_{1,sd}^h(t)\right)$$
$$= 1 - \left(1 - F_{0,sd}^h(t)\right)e^{-\lambda_{1,sd}^h t}. \tag{4.22}$$

Extending the computation of the cdf of the burst length at the second hop gives

$$F_{2,sd}^h(t) = 1 - \left(1 - F_{1,sd}^h(t)\right)\left(1 - H_{2,sd}^h(t)\right)$$
$$= 1 - \left(1 - F_{0,sd}^h(t)\right)e^{-(\lambda_{1,sd}^h + \lambda_{2,sd}^h)t}. \tag{4.23}$$

Generalizing the computation to kth hop gives

$$F_{k,sd}^h(t) = 1 - \left(1 - F_{k-1,sd}^h(t)\right)\left(e^{-\lambda_{k,sd}^h t}\right)$$
$$= 1 - \left(1 - F_{0,sd}^h(t)\right)e^{-\left(\sum_{i=1}^{k}\lambda_{i,sd}^h\right)t}. \tag{4.24}$$

To compute the packet loss probability from the cdf of the burst length at the ingress and egress nodes, the expected value of the term derived above needs to be computed and compared with the corresponding value at the ingress node which then gives the expected loss per burst. Let the expected length of the high priority burst at the end of k hops be denoted by $L_{k,sd}^h$. Depending on the burst length distribution at the ingress and the path length, Eq. 4.24 can be evaluated after substituting $F_0^h(t)$. For example, if the bursts are of fixed size T_0, the initial cdf of the burst length is given by

$$F_{0,sd}^h(t) = P(T \leq t) = \begin{cases} 1 & \text{if } t \geq T_0 \\ 0 & \text{if } t < T_0. \end{cases} \tag{4.25}$$

Substituting this into Eq. 4.24 and taking the expected value give the expected burst length after k hops to be

$$L_{k,sd}^h = \frac{1 - e^{-\sum_{i=1}^{k}\lambda_{i,sd}^h T_0}}{\sum_{i=1}^{k}\lambda_{i,sd}^h}. \tag{4.26}$$

Otherwise, if the initial burst length is exponentially distributed (with average $1/\mu$), so that

$$F_{0,sd}^h(t) = 1 - e^{-\mu t}, \tag{4.27}$$

using Eq. 4.24 and taking the expected value gives

$$L_{k,sd}^h = \frac{1}{\sum_{i=1}^{k}\lambda_{i,sd}^h + \mu}. \tag{4.28}$$

In this way, for each source–destination pair sd, with a fixed path length of l hops, the expected length of the burst lost for a high-priority burst is given by $LPB^h = L_{0,sd}^h - L_{l,sd}^h$ and the packet loss probability per sd pair for the high-priority bursts, PLP_{sd}^h, can be obtained by taking the ratio of LPB^h to the expected value of the initial length of the burst $L_{0,sd}^h$. Then the average packet loss probability for high-priority bursts in the entire network is computed by taking the weighted average of individual loss probabilities over all the source–destination pairs as

$$PLP^h = \sum_s \sum_d \frac{\lambda_{sd}^h}{\lambda^h} PLP_{sd}^h, \tag{4.29}$$

where λ^h is the total arrival rate of all the high-priority bursts into the network given by $\sum_s \sum_d \lambda_{sd}^h$.

Once the packet loss probability for high-priority bursts is computed, the next step is to compute the same for the low-priority bursts. Since a low-priority burst in service or that arrives later is preempted by a high-priority burst, the arrival rate of low-priority bursts depends on the utilization of the link by the high-priority bursts. On the first hop along a path, the load is that due to bursts from all the source–destination pairs but on all subsequent hops, the load is due to the bursts that are not blocked by the high-priority bursts at the previous hop. Thus the arrival rate at a hop k due to the low-priority bursts on a path between a source–destination pair sd, denoted by $\lambda_{k,sd}^l$, can be written as

$$\lambda_{k,sd}^l = \begin{cases} \lambda_{sd}^l, & \text{if } k = 0, \text{ i.e., before the first hop} \\ \lambda_{k-1,sd}^l (1 - v_{k-1}), & \text{for all hops } k \geq 1 \\ 0, & \text{if the } k\text{th hop is not on the path between } (s, d) \end{cases} \tag{4.30}$$

where v_i is the utilization of link i by the high-priority bursts. This can be expressed as

$$v_i = 1 - e^{\frac{-\lambda_i^h}{\overline{\mu}_i}}, \tag{4.31}$$

with $1/\overline{\mu}_i$ being the mean service time on a link i along the path. The average service time is computed by taking the weighted average of the link service time on all the links. That is, if $\mu_{i,sd} = 1/L_{i,sd}^h$ then

$$\frac{1}{\overline{\mu}_i} = \sum_{s,d:\lambda_{i,sd}^h > 0} \frac{\lambda_{i,sd}}{\lambda_i^h} \cdot \frac{1}{\mu_{i,sd}}. \tag{4.32}$$

Once the arrival rate of the low-priority bursts is obtained using Eq. 4.30, evaluation of the packet loss probability is similar to that for the case of high-priority bursts. Let the cdf of the initial burst length for the low-priority bursts between a pair of edge nodes (s, d) be $F_{0,sd}^l(t)$ and the same after k hops be $F_{k,sd}^l(t)$. Let $H_{k,sd}^l(t)$ be the cdf of the inter-arrival time of the bursts from the traffic between the same pair of nodes at the kth hop. To compute the packet loss probability of low-priority bursts, the cdf of the inter-arrival time of the bursts must include both the low- and high-priority bursts because both of these can lead to the loss of the bursts of low priority. Therefore,

$$H_{k,sd}^l(t) = 1 - e^{-(\lambda_{k,sd}^h + \lambda_{k,sd}^l)t}. \tag{4.33}$$

In this case, it is assumed that the low-priority burst being transmitted is segmented on the arrival of a burst of any priority. Note that analysis using this assumption does not hold for head-based segmentation but the analysis can be extended in a straightforward way for that case too. Considering that the length of the contending burst is always greater than the residual length of the original burst, the cdf of the burst length after the first hop is equal to the probability that the initial length is less than or equal to t or the next burst arrives in a time less than or equal to t. Expressing this mathematically gives

$$F_{1,sd}^l(t) = \nu_1 + (1 - \nu_1) \left[1 - \left(1 - F_{0,sd}^l(t) \right) \left(1 - H_{1,sd}^l(t) \right) \right]$$
$$= 1 - (1 - \nu_1) \left(1 - F_{0,sd}^l(t) \right) e^{-(\lambda_{1,sd}^h + \lambda_{1,sd}^l)t}. \tag{4.34}$$

Similarly, the cdf of the burst length after the second hop can be written as

$$F_{2,sd}^l(t) = \nu_2 + (1 - \nu_2) \left[1 - \left(1 - F_{1,sd}^l(t) \right) \left(1 - H_{2,sd}^l(t) \right) \right]$$
$$= 1 - (1 - \nu_2)(1 - \nu_1) \left(1 - F_{0,sd}^l(t) \right) e^{-(\lambda_{1,sd}^h + \lambda_{1,sd}^l + \lambda_{2,sd}^h + \lambda_{2,sd}^l)t}. \tag{4.35}$$

Generalizing the evaluation to k hops gives

$$F_{k,sd}^l(t) = \nu_k + (1 - \nu_k) \left[1 - \left(1 - F_{k-1,sd}^l(t) \right) \left(1 - H_{k,sd}^l(t) \right) \right]$$
$$= 1 - \prod_{i=1}^{k}(1 - \nu_i) \left(1 - F_{0,sd}^l(t) \right) e^{-(\sum_{j=1}^{k} \lambda_{j,sd}^h + \lambda_{j,sd}^l)t}. \tag{4.36}$$

Comparing the expected length of the burst after k hops with that at the source node gives the expected loss rate. Similar to the case of high-priority bursts, if the initial burst length is constant (say T_0), the expected length of the low-priority burst at the kth hop can be written as

$$L_{k,sd}^l = \frac{\prod_{i=1}^{k}(1 - \nu_i) \left(1 - e^{-\sum_{i=1}^{k}(\lambda_{i,sd}^h + \lambda_{i,sd}^l)T_0} \right)}{\sum_{i=1}^{k} \left(\lambda_{i,sd}^h + \lambda_{i,sd}^l \right)}, \tag{4.37}$$

whereas if the initial burst length is exponentially distributed (see Eq. 4.27), then

$$L^l_{k,sd} = \frac{\prod_{i=1}^{k}(1 - v_i)}{\sum_{j=1}^{k}\left(\lambda^h_{j,sd} + \lambda^l_{j,sd}\right) + \mu}. \tag{4.38}$$

The average packet loss probability of the low-priority bursts in the network is computed by taking the weighted average of the loss probabilities for each source–destination pair similar to the case of high-priority bursts.

The above analysis to compute the average packet loss probability due to prioritized segmentation for a system with two levels of priority can also be extended for multiple priority classes. However, the analysis has few shortcomings. First, it overestimates the packet loss probability by considering any two events of segmentation for the same pair of bursts on the same path to be independent. In practice, if two contending bursts follow the same route, then the original burst is segmented only at the first encounter. At the same time, the analysis does not consider the segmentation of the bursts at places other than the packet boundaries leading to a slight underestimation. Due to the burst segmentation, some packets may be received partially at the destination which must be accounted for in the computation of the loss. When the burst size is smaller, the loss due to partial packets is significant. Analysis of segmentation in such a case when the packet size is comparable to the burst size is dealt with in Section 4.2.1 which can be extended even for multiple priority classes. By accounting for the reduced load due to the bursts blocked at the upstream nodes the updated arrival rate of bursts can be used to improve the accuracy in computation of the packet loss probability.

4.2.4 Probabilistic Preemptive Burst Segmentation

All the burst segmentation techniques discussed so far have a deterministic policy to drop or segment one of the contending bursts. The scheme that is discussed in this section known as probabilistic preemptive burst segmentation (PPBS) [42] preempts and segments the bursts in a probabilistic fashion. By tuning the probability of burst segmentation it was shown that PPBS achieves proportional loss differentiation among multiple classes of traffic [42]. In the preemptive burst segmentation scheme discussed earlier, when a high-priority burst is in service, a contending low-priority burst gets preempted or segmented till the high-priority burst is serviced. In PPBS, a low-priority burst that contends with the high-priority burst in service can be either completely dropped or preempted (segmented). This is shown in Fig. 4.4 where the contending low-priority burst is dropped completely in case (a) while it is segmented in case (b), till the service of high-priority burst is completed. The decision to drop the burst or segment is taken using a probability value which differentiates PPBS from preemptive burst segmentation. A low-priority burst is completely dropped with a probability of p when it contends with a high-priority burst, whereas it is segmented with a probability of $1 - p$. Tuning the value of p allows to achieve the desired proportion of loss differentiation among different classes of traffic. Basically, for the same value of loss probability for the high-priority traffic, varying p gives different loss probabilities for the low-priority traffic. Similar to the case of preemption, PPBS can also be used when the two bursts have different priority levels. When both the bursts contending are of the same priority, PPBS scheme cannot be applied. Instead, the contending burst is dropped completely and the original burst is serviced.

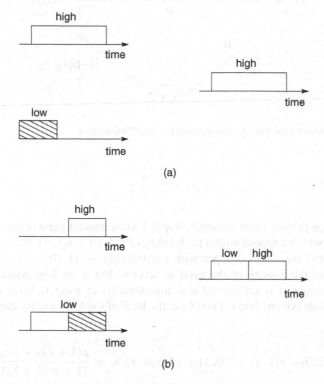

Fig. 4.4 Illustration of the probabilistic prioritized burst segmentation scheme

To understand how the variable p affects the loss probability, the PPBS scheme for a single wavelength is modeled using a CTMC whose state $s(i, j)$ is defined by two variables i and j which denote the number of high- and low-priority bursts, respectively, on a wavelength. Figure 4.5 shows the state transition diagram of the Markov chain for which the limiting state probabilities can be derived easily.

Let λ_h be the arrival rate of high-priority traffic, λ_l be the arrival rate of low-priority traffic, and λ be the cumulative rate. Let the service time of each class be distributed with a negative exponential distribution with a mean μ. For simplicity, assume that $\mu = 1$ and the time units are normalized by μ. The BLP of the high-priority class can be simply written as

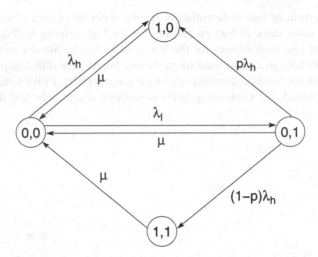

Fig. 4.5 Continuous-time Markov chain model for the PPBS scheme

$$PB_h = \frac{\lambda_h}{\lambda_h + 1} \qquad (4.39)$$

because a high-priority burst is simply dropped when another burst of the same class is in service which happens with a probability of $s(1, 0) + s(1, 1)$. If the new burst is a low-priority burst, it is dropped with a probability of $s(1, 0) + s(1, 1) + s(0, 1)$ irrespective of the priority of the burst in service. But if the low-priority burst is already in service, it is segmented with a probability of $p\lambda_h s(0, 1)/\lambda_l$ due to the arrival of a high-priority burst. Therefore, the BLP of the low-priority class is given by

$$PB_l = s(1, 0) + s(1, 1) + s(0, 1) + p\lambda_h s(0, 1)/\lambda_l = \frac{\rho(1 + \lambda_h) + p\lambda_h}{(1 + \rho)(1 + \lambda_h)}, \qquad (4.40)$$

where $\rho = \lambda/\mu$ is the total offered load.

The BLP derived above for a two-class system is for a link with only one wavelength. The analysis for multiple wavelengths in a link say w, $w \geq 1$ in number, can be done by using a multi-dimensional Markov chain with state space denoted by $s(i, j)$, $i = 0, 1, \ldots, w$ and $j = 0, 1, \ldots, w$ as defined earlier. The steady-state equations for the system can be written as

$$((\lambda_h + \lambda_l) + (i + j - 2))\, s(i - 1, j - 1) = \lambda_h s(i - 2, j - 1) + is(i, j - 1) + js(i - 1, j)$$
$$+ \lambda_l s(i - 1, j - 2),$$
$$(\lambda_h + k)s(i, w - i) = \lambda_h s(i - 1, w - i) + \lambda_l s(i, w - i - 1)$$
$$+ \lambda_h ps(i - 1, w - i - 1),$$
$$ws(k, 0) = \lambda_h s(w - 1, 0) + \lambda_h ps(w - 1, 1),$$
$$ws(w, w) = (1 - p)\lambda_h s(w - 1, w),$$

where the boundary conditions are $s(-1, j) = s(i, -1) = s(i, w + 1) = s(w + 1, j) = 0$. Though it is possible to solve the system of equations given above numerically, some results discussed in Section 3.9 can be used to analyze the BLP of a two-class system with w wavelengths.

First, note that in a classless system (without any priority among the bursts), the loss probability on a link with w wavelengths and offered load ρ is given by the Erlang's loss formula $B(\rho, w)$ (Eq. 3.1) which is simply the probability that all the wavelengths on the link are busy. This also happens to be the lower bound on the loss probability in a two-class system with w wavelengths. If the preemption probability, $p = 1$, the loss probability of the high-priority class is given by $B(\lambda_h, w)$ and the proportion of the time that a high-priority burst preempts the low-priority one is given by $B(\rho, w) - B(\lambda_h, w)$. Thus, the total low-priority bursts lost due to preemption in this time can be approximated by $\lambda_h(B(\rho, w) - B(\lambda_h, w))$ and the fraction of the low-priority bursts lost is $\lambda_h(B(\rho, w) - B(\lambda_h, w))/\lambda_l$. By the Markov property, preemption does not change the length of the burst or the occupancy of the wavelengths. Along with the loss due to preemption, a low-priority burst may also be lost when it finds all the wavelengths to be busy. Using ergodicity, the loss probability of the low-priority bursts is the sum of these two long-term averages. Hence, the upper bound on the loss probability of low-priority bursts is given by

$$PB_{l,\text{upp}} = \lambda_h(B(\rho, w) - B(\lambda_h, w))/\lambda_l + B(\rho, w) \qquad (4.41)$$

so that for a two-class system with w wavelengths and strict preemption ($p = 1$), the loss probability of low-priority bursts satisfies

$$PB_{l,\text{low}} = B(\rho, w) \leq PB_l \leq PB_{l,\text{upp}}. \qquad (4.42)$$

If the preemption probability $p < 1$, then the probability that a high-priority burst preempts a low priority is given by

$$P_p = p\lambda_h(B(\rho, w) - B(\lambda_h, w))/\lambda_l. \qquad (4.43)$$

Apart from this, a low-priority burst experiences a blocking similar to that with the Erlang B loss formula with a combined load of λ. With all these results derived so far, the BLP of high-priority traffic is given by $PB_h = B(\lambda_h, w)$ and that for the low-priority traffic is given by

$$\begin{aligned} PB_l &= B(\rho, w) + p\lambda_h(B(\rho, w) - B(\lambda_h, w))/\lambda_l \\ &= pPB_{l,\text{upp}} + (1 - p)PB_{l,\text{low}}, \end{aligned} \qquad (4.44)$$

where $PB_{l,\text{upp}}$ and $PB_{l,\text{low}}$ are defined by Eqs. 4.41 and 4.42, respectively. Further, according to the conservation law in OBS, the loss probability and the throughput averaged over all the classes remain constant regardless of the number of classes.

That is,

$$\lambda B(\rho, w) = \lambda_h P B_h + \lambda_l P B_{l,\text{upp}}. \tag{4.45}$$

So far in this section, traffic with only two priority levels is considered. The analysis can also be extended to a system with multiple classes of traffic. Let there be $M \geq 2$ classes of traffic on a link with w wavelengths. In such a system, any burst with a higher priority indicated by smaller index $j < i$ can preempt a low-priority burst with a probability p_{ji}, $j < i \in \{1, 2, \ldots, M-1\}$. The probability that the high-priority burst segments the low-priority one is correspondingly $1 - p_{ji}$. For a particular class i, $1 < i \leq M$, let N be the group of all classes with priority level higher than i. Applying the analysis for a two-class system with w wavelengths for the class i and the set N gives the loss of class i bursts due to preemption by the traffic from all the classes from set N. Since it is assumed that the arrival of bursts in any class of traffic follows a Poisson distribution, the sum of the arrival rate of bursts from all high priority classes that preempt bursts in class i traffic, is given by $\sum_{j=1}^{i-1} p_{ji}\lambda_j$. Using the same logic that leads to Eq. 4.41, the BLP of class i traffic due to preemption by the bursts from all the classes in the set N, can be written as

$$B\left(\sum_{j=1}^{i} \lambda_j, w\right) + \sum_{j=1}^{i-1} \frac{p_{ji}\lambda_j}{\lambda_i}\left(B\left(\sum_{j=1}^{i} \lambda_j, w\right) - B\left(\sum_{j=1}^{i-1} \lambda_j, w\right)\right), \quad i = 1, 2, \ldots, M.$$
$$\tag{4.46}$$

Let $\mathbf{p_i}$ be a column vector of size $M - 1$ with the first $i - 1$ components being the probabilities p_{ji} and the rest being zeros. Let Λ be a column vector of size $M - 1$ with the individual components being the input rates λ_j, $1 \leq j \leq M - 1$. The loss probability of the class i traffic as given in Eq. 4.46 can be written in a compact form as

$$BLP_i = R_i + \mathbf{p}_i^T \Lambda S_i, \tag{4.47}$$

where

$$R_i = B\left(\sum_{j=1}^{i} \lambda_j, w\right), \quad S_i = (R_i - R_{i-1})/\lambda_i. \tag{4.48}$$

4.2.5 Preemption Probability

In this section, the probability that a burst gets segmented also known as the *preemption probability* is computed and its relation with the burst length distribution is studied. As mentioned earlier, when two bursts contend at a core node, assuming that both of them have the same priority, either the head of the contending burst is segmented or the tail of the original burst is segmented. Usually, if the residual length of the original burst is greater than the contending burst, then the contending burst is either dropped or it is segmented till the original burst is served. If the residual length of the original burst is smaller than the incoming burst, then the original burst is segmented. This is also termed as preemption since the burst in service (original burst) is preempted by the contending burst. The probability that the contending burst wins or the probability that the original burst gets segmented is termed as preemption probability. Computing preemption probability is important because burst segmentation involves additional processing cost due to the generation of the trailer packet, rescheduling the bursts at the downstream nodes, and the switching overhead [32].

The main assumptions made for this analysis are as follows: the nodes do not have wavelength conversion capability, the switching time is negligible compared to the average burst length, the burst arrival process follows the Poisson distribution, and the input burst size distribution is determined by the time-based assembly algorithm and not influenced by segmentation. However, as shown through the analysis, the burst size distribution after segmentation changes since the average service time is shifted to larger values. Figure 4.6 shows the case when preemption occurs and introduces the notation used in the analysis. Assume that the arrival time and length of the first burst on a free wavelength (shown as burst 0 in the figure) is denoted by (t_0, l_0). Let (t_i, l_i), $i = 1, 2, \ldots, n$ denote the arrival time and the length of the bursts that arrive during the time when the first burst (burst 0) is being serviced such that $t_0 < t_i < t_0 + l_0$ for all $i = 1, 2, \ldots, n$. According to the condition for preemption mentioned earlier, an ith burst preempts the 0th burst if

$$l_i > l_0 - \sum_{j=1}^{i}(t_j - t_{j-1}) = l_0 - (t_i - t_0) \qquad (4.49)$$

and $l_k \leq l_0 - \sum_{j=1}^{k}(t_j - t_{j-1})$, $k = 1, 2, \ldots, i - 1$. Without the loss of generality, assume that the nth burst wins the contention and the remaining $n - 1$ bursts are dropped. Let the arrival time and length of the preempting burst, i.e., (t_n, l_n), be denoted as (t', l'). Intuitively the preempting burst has a larger probability of having a larger service time compared to the original burst so that the service time distribution of the burst winning contention, i.e., (t', l'), is shifted to larger values.

Let F be the cdf of the service time and L be the random variable with a cdf F. Let S_n indicate the set of all possible n arrivals and p_n be the probability of n arrivals in $(t_0, t_0 + l_0)$. Note that p_n is a Poisson measure and $\sum_{n=0}^{\infty} \sum_{S_n} P(S_n)p_n = 1$ because preemption is assumed to occur definitely. Therefore, it is straightforward to see that

$$P(l' > x) = \sum_{n=0}^{\infty} \sum_{S_n} P(l_n > x|S_n)P(S_n)p_n. \qquad (4.50)$$

But $P(l_n > x|S_n)$ can be written as

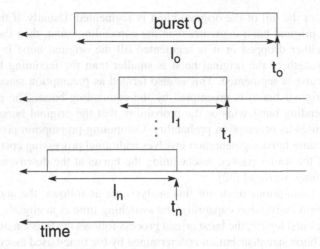

Fig. 4.6 Illustration of burst preemption

$$P(l_n > x | S_n) = P\left(L > x | L > l_0 - \sum_{j=1}^{n}(t_j - t_{j-1})\right)$$

$$> P(L > x | L > 0) = P(l_0 > x).   \tag{4.51}$$

Substituting Eq. 4.51 into Eq. 4.50 and using the condition $\sum_{n=0}^{\infty}\sum_{S_n}P(S_n)p_n = 1$ give $P(l' > x) > P(l_0 > x)$ which proves that the service time of the preempting burst is shifted to larger values.

Since this analysis does not make any assumptions on the burst size distribution, the result obtained in Eq. 4.51 holds good for any burst length distribution. The fact that there is a higher probability for a burst that wins contention to have a larger length signifies that there is a higher probability for the contending burst to win the contention in subsequent attempts. In other words, the burst that is segmented on the first attempt has a higher chance of getting segmented on subsequent attempts because its length would be shorter than that of the contending bursts in each subsequent contention. Therefore, the preemption probability has a maximum value for the first burst in a busy period. This can be expressed as

$$P(L > R) = \int_0^{\infty} P(L > x) dF_R(x),   \tag{4.52}$$

where R is the random variable denoting the residual life of the service time with density $f_R(x)$ given by [25]

$$f_R(x) = \frac{P(L > x)}{E[X]}, \quad x > 0.   \tag{4.53}$$

For exponentially distributed service times the preemption probability can be obtained as

$$P(L > R) = \int_0^\infty e^{-\lambda x} \lambda e^{-\lambda x} dx = 1/2. \tag{4.54}$$

Similarly, for service times with the Pareto distribution

$$P(L > x) = \begin{cases} 1, & x < K \\ K^\alpha x^{-\alpha}, & x \geq K \end{cases}, \tag{4.55}$$

so that using Eq. 4.53 the residual life of service time is given by

$$P(R > y) = \begin{cases} \frac{(\alpha-1)(K-y)+K}{\alpha K}, & 0 \leq y \leq K \\ \frac{K^{(\alpha-1)} y^{(1-\alpha)}}{\alpha}, & y > K \end{cases} \tag{4.56}$$

and the preemption probability is given by

$$P(L > R) = \frac{2(\alpha - 1)}{2(\alpha - 1) + 1}. \tag{4.57}$$

Comparing the preemption probability for burst length with exponential distribution and Pareto distribution (Eqs. 4.54 and 4.57) shows that the preemption probability is influenced by the burst length distribution. For Pareto-distributed burst sizes it can be seen that the preemption probability depends on the shape factor α while it depends on the rate of the arrival of the bursts for exponentially distributed burst sizes. For the same load, the preemption probability is influenced by the burst size distribution which also depends on the assembly algorithm used at the edge node. As seen for the case of Pareto distribution, as the variability in the length decreases (indicated by increasing α or coefficient of variation in Gaussian distribution) preemption probability increases. An interesting conclusion made from this analysis is that the preempting burst tends to have a higher length while the segmented burst has a larger chance of getting segmented again. So the preemption probability is largely controlled by the distribution of the size of the first burst in a busy period which is significantly influenced by the burst assembly algorithm.

4.3 Wavelength Preemption

Though it was shown that burst segmentation improves the performance of OBS networks by reducing the packet loss probability, it causes increased complexity at the core nodes. It also involves modification of the architecture of the core node to support generation of trailer packets and to identify the packet boundaries. Apart from this there may be some additional mechanisms required to assemble the bursts considering the priorities of the packets which leads to an increased assembly delay at the ingress nodes. In a network supporting multiple classes of traffic, prioritized segmentation is complex to implement due to the number of combinations that arises with priorities of bursts as seen in Section 4.2.3. To overcome these shortcomings, strict preemption of bursts where one of the bursts is dropped instead of being segmented is considered. Wavelength preemption is proposed in [37, 39] along with a scheme for differentiated bandwidth allocation to multiple classes of traffic to provide a desired level of service differentiation. In the wavelength preemption technique, a high-priority burst that contends with a low-priority burst is serviced not by preemption of the low-priority burst but is rescheduled to deny the wavelength reservation to the low-priority burst. Since the wavelength is reserved when the BHP is processed, rescheduling the wavelengths in order to transmit high-priority bursts allows the low-priority bursts to be dropped with a deterministic rate. Assuming that the probability with which a high-priority burst is rescheduled to preempt a low-priority burst is determined by another scheme like the preemption-based bandwidth allocation in [39], theoretical analysis of the BLP of the low-priority bursts shows that the loss rate of each class of the bursts can be controlled at a desired level.

In this section, the BLP due to a generic wavelength preemption technique at the core node with w wavelengths on the outgoing link is analyzed initially for two classes of traffic and then for the case of multiple classes [39]. The number of wavelengths used by the bursts from a class of traffic is modeled as a CTMC and the BLP of each class of traffic is derived. Let the bursts of a class i, $i = 1, 2, \ldots, N$, arrive according to a Poisson distribution with a mean rate of λ_i with their length being exponentially distributed with a mean $1/\mu$. The state of the CTMC is represented by the tuple $s = (a_1, a_2, \ldots, a_N)$ where a_n is the number of wavelengths used by the traffic of class n. When all the wavelengths are busy the state is represented by $\phi \in \{s | \sum_{j=1}^{N} a_j = w\}$. Let $p_{ij}(\phi)$ denote the probability that a burst of class i preempts a burst of class j when the link is in state ϕ which can be set to any value in this analysis.

When there is a single wavelength and two classes say, l and h with h being the class with a higher priority, the state of the link $s = (a_l, a_h)$ indicates the number of bursts of classes l and h in transmission, respectively. The CTMC representation is shown in Fig. 4.7 where the state $(0, 0)$ indicates that the link is idle and the states $(0, 1)$ and $(1, 0)$ indicate that the wavelength is transmitting a burst of class h and class l, respectively. For example, when the link is in the state $(1, 0)$ a new burst of higher priority may preempt the burst of class l with a probability $p_{hl}(1, 0)$ and the state changes to $(0, 1)$. In state $(0, 1)$, the arrival of either a low-priority or a high-priority burst does not change the state because the burst of class h in service cannot be preempted.

Let the probability of being in a state (i, j) be denoted by $P(i, j)$. Using the state transition diagram in Fig. 4.7, the following equations can be obtained:

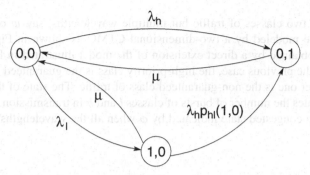

Fig. 4.7 Continuous-time Markov chain model for wavelength preemption scheme on a link with a single wavelength serving two classes of traffic

$$\lambda_h P(0, 0) + p_{hl}(1, 0)\lambda_h P(1, 0) = \mu P(0, 1),$$
$$\lambda_l P(0, 0) = \mu P(1, 0) + p_{hl}(1, 0)\lambda_h P(1, 0),$$
$$P(0, 0) + P(0, 1) + P(1, 0) = 1, \tag{4.58}$$

solving which the steady-state probabilities of different states can be obtained as

$$P(0, 0) = \frac{\mu}{\lambda_l + \lambda_h + \mu},$$
$$P(0, 1) = \frac{\lambda_h(\mu + p_{hl}(1, 0)\lambda_l + p_{hl}(1, 0)\lambda_h)}{(\lambda_l + \lambda_h + \mu)(\mu + p_{hl}(1, 0)\lambda_h)},$$
$$P(1, 0) = \frac{\lambda_l \mu}{(\lambda_l + \lambda_h + \mu)(\mu + p_{hl}(1, 0)\lambda_h)}. \tag{4.59}$$

It can be seen from Fig. 4.7 that the newly arriving high-priority bursts are lost whenever they fail to preempt the ongoing transmission on a link (i.e., either in state $(0, 1)$ or state $(1, 0)$). However, the low-priority bursts are lost when the link is in either state $(0, 1)$ or state $(1, 0)$ so that they are preempted by the high-priority bursts. From these conditions, the BLP for the two classes of bursts can be obtained by taking the ratio of the number of bursts that are lost to the total number of arrivals on the wavelength. Therefore, the BLP for the low-priority and the high-priority bursts can be written as

$$PB_l = \frac{\lambda_l(P(0, 1) + P(1, 0)) + \lambda_h p_{hl}(1, 0)P(1, 0)}{\lambda_l(P(0, 0) + P(1, 0) + P(0, 1))}$$
$$= (P(1, 0) + P(0, 1)) + \frac{\lambda_h}{\lambda_l} p_{hl}(1, 0)P(1, 0),$$
$$PB_h = \frac{\lambda_h P(0, 1) + \lambda_h P(1, 0)(1 - p_{hl}(1, 0))}{\lambda_h(P(0, 0) + P(1, 0) + P(0, 1))}$$
$$= P(0, 1) + (1 - p_{hl}(1, 0))P(1, 0). \tag{4.60}$$

With only two classes of traffic but multiple wavelengths, say w of them, the system can be modeled by a two-dimensional CTMC as shown in Fig. 4.8 [38]. This can be obtained by a direct extension of the model illustrated in Fig. 4.7. As described in the previous case, the high-priority class is the guaranteed traffic class while the other one is the non-guaranteed class of traffic. The state of the link $s = (a_l, a_h)$ indicates the number of bursts of classes l and h in transmission. The link is said to be in a congested state indicated by ϕ when all the wavelengths on the link are busy.

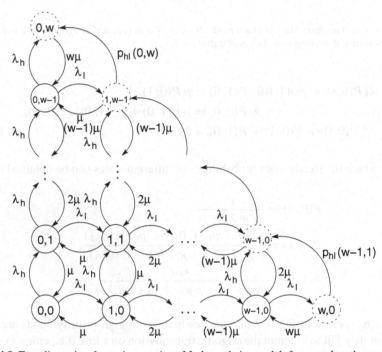

Fig. 4.8 Two-dimensional continuous-time Markov chain model for wavelength preemption scheme on a link with w wavelength serving two classes of traffic

From the Markov chain in Fig. 4.8 the following rate equations can be written:

$$(\lambda_l + \lambda_h + (i + j)\mu)\, P(i, j) = \lambda_l \delta(w - i)P(i - 1, j) + (i + 1)P(i + 1, j)\mu$$
$$+ \lambda_h \delta(w - j)P(i, j - 1) + (j + 1)$$
$$P(i, j + 1)\mu \ (0 \le i + j < w), \qquad (4.61)$$

$$(\lambda_h \delta(j)p_{hl}(i - 1, j + 1) + w\mu)\, P(i, j) = \lambda_l \delta(j)P(i - 1, j)$$
$$+ \lambda_h \delta(i)P(i, j - 1) \quad (i + j = w), \qquad (4.62)$$

$$\sum_{i=0}^{w}\sum_{j=0}^{w-1} P(i, j) = 1, \qquad (4.63)$$

where $\delta(i)$ is the unit step function defined as

$$\delta(i) = \begin{cases} 0 & \text{for } i = w \\ 1 & \text{for } i < w. \end{cases} \qquad (4.64)$$

It can be seen from Fig. 4.8 that whenever a low-priority burst finds the link busy (state ϕ) or it is preempted by another high-priority burst, there is a loss of low-priority traffic. On the other hand, the high-priority traffic is lost either when the state is $(0, w)$, i.e., all the wavelengths are serving a high-priority burst and therefore preemption is not possible or preemption fails in any other state. The probability of a failed preemption attempt is given by $1 - p_{hl}(s)$ in a state s. Therefore, the BLP for the high-priority and low-priority traffic can be written as [38]

$$PB_h = P(0, w) + \sum_{i=1}^{w}(1 - p_{hl}(i-1, w-i+1))P(i, w-i),$$

$$PB_l = \sum_{i=1}^{w} P(i, w-i) + \frac{\lambda_h}{\lambda_l} \sum_{i=1}^{w} p_{hl}(i-1, w-i+1)P(i, w-1). \quad (4.65)$$

The analysis can also be extended on similar lines in the case of multiple classes (say N classes) of traffic with state-dependent preemption probability. The expression for the BLP given here is for a generic preemption algorithm. For a specific preemption scheme, the value of $p_{ij}(s)$ might vary according to the way preemption is defined among bursts of multiple classes. The reader is referred to [38, 39] for some sample wavelength preemption mechanisms. The analysis provided in this section assumes that class i traffic belongs to the set of non-guaranteed traffic classes. For a guaranteed class of traffic, the BLP can be computed from the first case defined below. The BLP of a newly arrived burst of class i can be written as

$$PB_i = \sum_{s_1} P(s_1) + \sum_{s_2} \left(1 - p_{ij}(s_2^{ij})\right) P(s_2) + \sum_{s_3} \frac{\lambda_j}{\lambda_i} p_{ji}(s_3^{ji} P(s_3))$$

$$= PB_i(s_1) + PB_i(s_2) + PB_i(s_3), \qquad (4.66)$$

where if state $s_1 = (a_1, a_2, \ldots, a_N)$, then $s_2^{ij} = (a_1, a_2, \ldots, a_{j-1}, a_j - 1, a_{j+1}, \ldots, a_{i-1}, a_i + 1, a_{i+1}, \ldots, a_N)$, $(j < i)$ and $s_3^{ji} = (a_1, a_2, \ldots, a_{i-1}, a_i - 1, a_{i+1}, \ldots, a_{j-1}, a_j + 1, a_{j+1}, \ldots, a_N)$, $(j > i)$. The three components of the BLP in Eq. 4.66 can be interpreted as follows:

1. $PB_i(s_1)$ is the probability that a burst arrives in state ϕ and all the wavelengths are reserved by the bursts of classes with larger or equal priority than i.
2. $PB_i(s_2)$ is the probability that a new burst that arrives in state ϕ cannot preempt the burst in transmission.

3. $PB_i(s_3)$ is the probability that a burst in transmission can be preempted by a burst of higher priority in the state ϕ.

The set of states and the corresponding loss probabilities in those states vary according to the class of traffic as follows:

1. When $i = N$ (highest priority class),

$$s_1 \in \{\phi | a_k = 0; k = 1, 2, \ldots, N - 1\},$$
$$s_2 \in \{\phi | a_N \neq w\},$$
$$PB_i(s_3) = 0. \tag{4.67}$$

2. When $1 < i < N$ (any other class),

$$s_1 \in \{\phi | a_k = 0; k = 1, 2, \ldots, i - 1\},$$
$$s_2 \in \{\phi | a_i \neq w; k = i, i + 1, \ldots, N\},$$
$$s_3 \in \{\phi | a_i \neq 0 \cap a_k \neq w; k = i + 1, i + 2, \ldots, N\}. \tag{4.68}$$

3. When $i = 1$ (non-guaranteed class),

$$s_1 \in \{\phi\},$$
$$PB_i(s_2) = 0,$$
$$s_3 \in \{\phi | a_1 \neq 0 \cap a_k \neq w; k = 2, 3, \ldots, N\}. \tag{4.69}$$

4.4 Deflection Routing

Deflection routing is another technique used in OBS networks for contention resolution. It is demonstrated through simulations that deflection routing is an effective contention resolution technique in OBS networks, especially at low and medium load conditions [24, 47, 48]. Deflection routing was also used earlier as a contention resolution technique in optical mesh networks with regular topologies [2, 4]. Various analytical models were proposed to study the performance of deflection routing in regular topologies. Deflection routing was also studied in buffered networks for a specific network topology or a traffic model [10, 12]. Through analysis, unslotted network was found to be a better approach to implement deflection routing than slotted network [5, 6].

The concept of deflection routing is pretty simple but there are some difficulties associated with its implementation when the nodes do not have buffers. In deflection routing, instead of dropping the contending bursts at an output port due to lack of wavelengths, they are routed through an alternate path to the destination. Although the bursts deflected on the alternate path may pass through a longer path, they are ultimately delivered to the destination reducing the BLP. When using deflection routing, bursts are dropped only if all the alternate routes are also busy. It is easier to implement deflection routing in circuit-switched networks or packet-switched networks with buffers. Usually, a pair of paths to the destination is computed at each node ahead of time. When the traffic needs to be deflected, the next best hop is used to reach the destination. In networks with buffers, if all the ports on the primary outgoing link are busy, traffic is buffered for some time before it is deflected through an alternate route.

Selection of the alternate route for deflected bursts is challenging because the deflected bursts increase the load in the other paths and there might be routing loops if the protocol is not designed carefully. There are several mechanisms that determine the choice of optimal path to destination for the deflected bursts [15, 28]. Based on the number of hops to the destination along both the primary and the deflected routes, it can be decided whether to deflect the bursts or drop them at the core node [28]. In some cases deflection routing from a node might increase the path length significantly compared to retransmitting the burst from the source node. To optimize the use of network resources and improve the BLP in deflection routing, a choice can be made about the route to be followed by the deflected bursts. With a centralized scheme to determine the minimum delay and blocking along the various paths to the destination, a list of alternate routes can be maintained at each ingress node which is used to update the routing tables at the core nodes. In a distributed approach, when there is a contention, the core node should dynamically choose the lightpath which minimizes the BLP. In selecting the lightpath bursts with higher tolerance to loss can be preempted by the bursts which have lower tolerance to loss [27]. It is obvious that deflection routing requires FDLs or additional offset time to delay the burst while updating the routing tables at the core nodes. An offset time-based on-demand deflection routing protocol proposed in [28] gives a linear programming formulation for the problem of computing alternate routes in the network which minimize both the delay and the blocking probability due to deflection routing. Along with this, judicious choice of whether to deflect the burst or drop it and retransmit later is shown to improve the performance of the network significantly compared to the case of indiscriminate deflection routing.

In OBS networks apart from the fact that core nodes may not have optical buffers which makes it difficult to delay the deflection of the bursts, there may be insufficient offset time due to deflection. Since the offset time is pre-computed for a specific number of hops, if the data burst must follow the BHP even after deflection, offset time must be kept sufficient enough to account for possible deflections in the future. Apart from this there might be an increase in the processing time of a BHP to reserve the wavelength on the outgoing links. As seen by the models in this section, even though deflection routing reduces the BLP in OBS networks, it also destabilizes the network in high load conditions. Since deflection routing increases the load at the alternate ports, analytical models that study the exact impact of deflection routing on the BLP in the network are necessary. This section studies models for BLP in OBS networks with deflection routing and discusses methods to avoid the instability in the network in the presence of deflections.

4.4.1 Modeling Deflection Routing at a Single Port

In JET-based OBS networks, the offset time is computed exactly sufficient for a path of known length between the source and the destination. At each hop as the BHP is processed the remaining offset time decreases so that at the destination, the burst arrives just after the BHP is processed. Due to this exact determination of the offset time, enabling deflection routing over a path of longer length causes insufficient offset time at the downstream nodes. If the core nodes have free FDLs, the data burst can be delayed for additional time. To ensure that bursts are not blocked due to insufficient offset time arising out of deflection routing, either extra offset time is provided which accounts for any possible deflections along the path or the deflected bursts are provided with buffers to adjust the offset time [20]. Both the methods to avoid burst loss due to insufficient offset time have their own advantages and disadvantages. By providing additional offset time so that even if there are deflections along the path, there is enough gap between the BHP and data burst, the priority of the burst increases. So the extra offset time cannot be increased to allow multiple deflections along a path or deflection through a significantly longer path. It also affects the delay performance of the traffic which is deflected. If the extra offset time is provided for future possible deflections and there are no deflections due to possibly light load, there is wastage of bandwidth and unnecessary delay of the bursts. Therefore, providing FDLs along the path and buffering the data bursts so that there is sufficient offset time is a better option. Therefore, it is difficult to implement deflection routing without causing instability in the network unless buffers are used [5, 20].

This section provides a basic model for the blocking probability of an OBS core node using JET protocol and deflection routing. Assume that FDLs are provided at each node either in a share-per-port or share-per-node configuration [17]. Let there be w wavelengths on each link and n FDLs at each node among which n_d are designated to the deflected bursts and the remaining ones for the non-deflected bursts. The number of virtual FDLs for deflected bursts is simply obtained by multiplying the number of physical FDLs with the number of wavelengths, i.e., $v_d = n_d w$. For simplicity, consider only one output port of a switch with a single output link where both the deflected and the non-deflected bursts arrive following a Poisson distribution with mean, λ_d and λ_f, respectively. Both the bursts are serviced with same rate $\mu = 1/L$ where the average burst length is exponentially distributed with mean L.

To maintain sufficient offset time between the BHP and the data burst even after deflection, each deflected burst is delayed by an FDL before it is served. The delay provided by the FDL is at least equal to the extra offset time required so that the burst does not overtake the BHP. The extra offset time can be computed by looking at the additional hops the burst has to travel due to deflection. This process at the output port is modeled as a Markovian system with two stages as shown in Fig. 4.9.

The first stage is an $M/M/w/w$ queue which models the behavior of the deflected bursts in FDLs. In the figure, $D_i, i = 1, 2, \ldots, v_d$, denotes the ith virtual

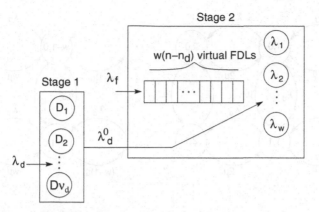

Fig. 4.9 Two-stage Markovian model for output port

FDL dedicated to the deflected bursts with an average service rate of $\mu_d = 1/(\delta h)$, where δ is the maximum processing time of the BHP at a node and h is the additional number of hops to the destination on an average. The loss probability at this stage can be written using Eq. 3.1 as

$$PB_1 = \frac{(\lambda_d/\mu_d)^{v_d}/v_d!}{\sum_{k=0}^{v_d}(\lambda_d/\mu_d)^k/k!}. \tag{4.70}$$

Since only deflected bursts are provided with the buffers, the blocking probability at this stage is dependent on the arrival rate of the deflected bursts alone. After this stage, both the deflected and the non-deflected bursts are directed to the output wavelengths. Assume that the departure of the bursts from stage 1 also follows Poisson distribution with a mean rate $\lambda_d^o = \lambda_d(1 - PB_1)$. In stage 2, both the deflected and the non-deflected bursts compete for the wavelengths. If all the wavelengths are busy, the bursts are delayed temporarily by the FDLs. To give the deflected bursts a higher priority in this stage, it is assumed that stage 2 provides a preemptive priority service with the deflected bursts preempting the non-deflected bursts. The non-deflected bursts can be buffered temporarily when preempted by the deflected bursts. Such a setting can be modeled by the Markov chain shown in Fig. 4.10.

Each state in Fig. 4.10 is represented by a tuple (i, j); $0 \leq i \leq w$ and $0 \leq j \leq w + (n - n_d).w$ are the number of deflected and non-deflected bursts in stage 2, respectively. It is not difficult to compute the number of states in the Markov chain as

$$n_s = \frac{(2(n - n_d)w + w + 2)(w + 1)}{2} = (w + 1)\left((n - n_d)w + \frac{w}{2} + 1\right). \tag{4.71}$$

Let π_{ij} be the steady-state probability that the system is in state (i, j), so that the stationary probabilities can be obtained from a system of difference equations given

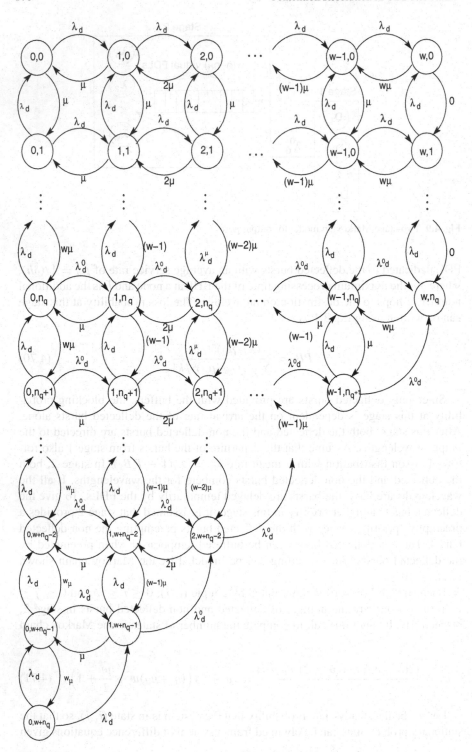

Fig. 4.10 State transition diagram of Markov chain-based model for stage 2

below:

$$0 = -(\lambda_d^o + \lambda_f)\pi_{00} + \mu\pi_{10} + \mu\pi_{01},$$

$$0 = -(\lambda_d^o + w\mu)\pi_{0,w+(n-n_d)w} + \lambda_f\pi_{0,w+(n-n_d)w-1},$$

$$0 = -(\lambda_f + w\mu)\pi_{w,0} + \lambda_d^o\pi_{w-1,0},$$

$$0 = -w\mu\pi_{w,w(n-n_d)} + \lambda_d^o(\pi_{w-1,w(n-n_d)} + \pi_{w,w(n-n_d)+1}) + \lambda_f\pi_{w,w(n-n_d)},$$

$$0 = -(\lambda_d^o + \lambda_f + k\mu)\pi_{k0} + \lambda_d^o\pi_{k-1,0} + (k+1)\mu\pi_{k+1,0} + \mu\pi_{k1},$$
$$1 \le k \le w-1,$$

$$0 = -(\lambda_d^o + \lambda_f + \min\{w,k\}\mu)\pi_{0k} + \lambda_f\pi_{0,k-1} + \mu\pi_{1k} + \min\{w,k+1\}\mu\pi_{0,k+1},$$
$$1 \le k \le w + (n - n_d)w,$$

$$0 = -(\lambda_d^o + k\mu + (w-k)\mu)\pi_{k,w+(n-n_d)w-k} + \lambda_d^o\pi_{k-1,w+(n-n_d)w-k}$$
$$+\lambda_d^o\pi_{k-1,w+(n-n_d)w-k+1} + \lambda_f\pi_{k,w+(n-n_d)w-k-1}, 1 \le k \le w-1,$$

$$0 = -(\lambda_f + w\mu)\pi_{w,k} + \lambda_d^o\pi_{w-1,k} + \lambda_f\pi_{w,k-1}, 1 \le k \le w(n-n_d) - 1,$$

$$0 = -(\lambda_d^o + \lambda_f + k\mu + \min\{w-k,l\}\mu)\pi_{k,l} + \lambda_d^o\pi_{k-1,l} + \lambda_f\pi_{k,l-1}$$
$$+(k+1)\mu\pi_{k+1,l} + \min\{w-k,l+1\}\mu\pi_{k,l+1}, 1 \le k \le w-1,$$
$$1 \le l \le w + (n-n_d)w - 2, \quad 2 \le k+l \le w + (n-n_d)w - 1,$$

$$1 = \sum \pi_{k,l}, \quad 0 \le k \le w, \quad 0 \le l \le w + (n-n_d)w,$$
$$0 \le k + l \le w + (n-n_d)w. \tag{4.72}$$

Let $\bar{x}$ denote the row vector $[0 \ \ 0 \ \ \cdots \ \ 0 \ \ 1]$ of dimension n_s and $\bar{\pi} = \begin{bmatrix} \pi_{00} & \pi_{01} & \pi_{10} & \cdots & \pi_{w,w(n-n_d)} \end{bmatrix}$ with $n_s + 1$ elements. Equation 4.72 can be expressed in a compact form as

$$\bar{x} = \bar{\pi}\begin{bmatrix} -(\lambda_d^o + \lambda_f) & 1 \\ \mu & 1 \\ \mu & 1 \\ 0 & \cdots & \vdots \\ \vdots & 1 \\ 0 & 1 \end{bmatrix}_{n_s \times n_s + 1}. \tag{4.73}$$

Solving the system of equations, the loss probability of stage 2 can be obtained as

$$PB_2 = \frac{\lambda_d^o}{\lambda_d^o + \lambda_f} \sum_{k=0}^{w(n-n_d)} \pi_{w,k} + \frac{\lambda_f}{\lambda_d^o + \lambda_f} \sum_{j=0}^{w} \pi_{j,w(n-n_d)-j}. \tag{4.74}$$

The total loss probability at the output port with deflection routing can be obtained by combining PB_1 and PB_2 considering the relative occupancy of each stage as

$$PB = \frac{\lambda_d}{\lambda_d + \lambda_f} PB_1 + \frac{\lambda_d^0 + \lambda_f}{\lambda_d + \lambda_f} PB_2. \qquad (4.75)$$

4.4.2 Probabilistic Deflection Routing

The above analysis considers an OBS node with a single port and full wavelength conversion capability. This section analyzes the BLP at a node with multiple ports and either with or without wavelength conversion capability [9]. As mentioned earlier, it is essential to limit the number of deflections or the load due to the deflections at a node because they increase contention losses especially under heavy load. Alternately, deflection routing can be invoked with a probability $0 < p < 1$ to limit the load due to deflections [9]. In low load situation, p can be set close to 1 so that every burst about to be dropped is deflected, whereas under heavy load deflection can be disabled ($p = 0$) or p can be close to 0. By controlling the value of p according to the load, instability in the network can be avoided. Also as inferred from the analysis in the previous section, the complexity of the model for deflection routing is high for nodes with multiple ports and under heavy load. By limiting the load with value of p, the load can always be maintained at a reasonable level so that the model proposed in this section is tractable.

To analyze the BLP at a node with multiple ports and probabilistic deflection routing, multi-dimensional Markov chains are used. Assume that the probability of deflection is constant at all ports. Further, a burst that arrives at a port is deflected to another port with a uniform probability. For simpler analysis, assume that the traffic intensity and BLP at a port are not affected by the deflected bursts from the other ports. This means that the deflected bursts from the other ports are not considered while computing the BLP at a port. With this assumption, approximate model is obtained for a node with deflection routing for either with or without wavelength conversion. The approximate model is good enough when the load is low so that the deflection events are not quite frequent.

4.4.2.1 Approximate Model with Wavelength Conversion

Consider a node with full wavelength conversion capability and w wavelengths per port. If the state of a node is modeled as the number of wavelengths occupied by the bursts, the node can be modeled by the Markov chain shown in Fig. 4.11. The node is in state i when there are i bursts being transmitted simultaneously. The arrival rate and the service rate of bursts in state i are denoted by λ_i and μ_i, respectively. Let the steady-state probability of state i be denoted by π_i.

For any state $0 < i < w$, the steady-state equations for the Markov chain shown in Fig. 4.11 can be written as

$$(\lambda_i + \mu_i)\pi_i - \lambda_{i-1}\pi_{i-1} - \mu_{i+1}\pi_{i+1} = 0. \qquad (4.76)$$

For $i = 0$

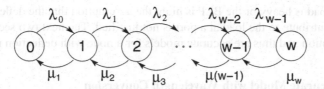

Fig. 4.11 State transitions in approximate model for node with wavelength conversion

$$\lambda_0 \pi_0 - \mu_1 \pi_1 = 0 \tag{4.77}$$

and for $i = w$

$$\mu_w \pi_w - \lambda_{w-1} \pi_{w-1} = 0. \tag{4.78}$$

The transition probabilities also satisfy $\sum_{i=0}^{w} \pi_i = 1$.

To compute the probability of a burst being blocked, note that without deflection routing it is π_m but with deflection routing, a burst is dropped only if the system is in state w (i.e., all the wavelengths are occupied) and with a probability p deflection routing is not used, or even if deflection routing is used there are no free wavelengths. This can be expressed in the form of an equation as follows:

$$PB = \pi_w (1 - p) + \pi_w \sum_{k=1}^{w-1} p_k . BLP_k, \tag{4.79}$$

where BLP_k is the BLP on kth port. Assuming that the probability of deflection and the loss rate are same for all the ports, the BLP can be written as

$$PB = \pi_w [(1 - p) + p . BLP], \tag{4.80}$$

which can be computed from the Erlang B formula if the burst arrival is assumed to be a Poisson process.

4.4.2.2 Approximate Model Without Wavelength Conversion

If there is no wavelength conversion capability, the burst arriving on a wavelength can be deflected to an output port only on the same wavelength. Since it is assumed that the bursts arrive independent of each other at an output port, the model for the case of no wavelength conversion is simplified to a two-state Markov chain where each wavelength is modeled independently. The state of the wavelength is 0 if it is free and 1 if it is occupied so that a new burst is lost. The BLP is simply obtained by replacing π_w by π_1 in Eq. 4.80 and by computing the BLP using the blocking probability in an $M/M/1/1$ queue.

As noted earlier, the approximate model given here ignores the effect of deflected traffic from the other ports and thus is not suitable under heavy load conditions.

When the load is heavy or the BLP is high, the assumption that the deflected traffic does not contribute to the load at a port is not justified. The next two sections relax this assumption and thus are accurate models for a node with deflection routing.

4.4.2.3 Accurate Model with Wavelength Conversion

To account for the effect of deflected bursts, the state of the system should not only include the number of reserved wavelengths in the port under consideration but also in the other ports. The accurate model is illustrated in Fig. 4.12 only for the case of two ports (say 1 and 2) but can be extended easily for any number of ports.

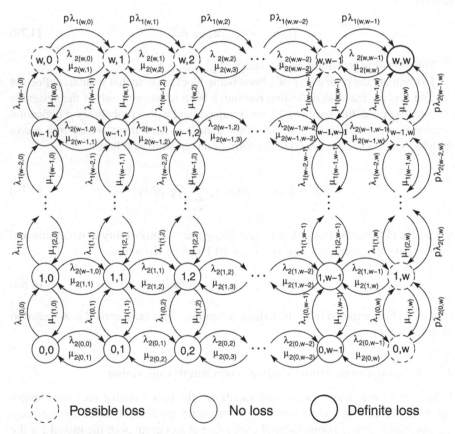

Fig. 4.12 State transitions in the accurate model for a node with wavelength conversion

For each port the state indicates the number of wavelengths reserved. State (i, j) indicates that the number of wavelengths being used in ports 1 and 2 are i and j, respectively. Let $\lambda_1(i, j)$ be the rate at which bursts arrive at port 1 when the system is in state (i, j) and $\lambda_2(i, j)$ is defined in a similar manner for port 2. Let $\mu_1(i, j)$ and $\mu_2(i, j)$ be the service rate of bursts at port 1 and port 2, respectively. Assume that the arrival rates are independent of the state and are denoted simply by $\lambda_1(i, j) = \lambda_1$ and $\lambda_2(i, j) = \lambda_2$ while the service rates are given by $\mu_1(i, j) = i\mu$ and $\mu_2(i, j) = j\mu$. With these definitions, the state transitions are similar to those defined in the approximate model for the case of $0 \le i, j < w$. The difference is only that there is a Markov chain for each port describing the transitions. For the state $i = w$, if a new burst arrives at port 1, it might be deflected to port 2 only if $j < w$, i.e., if there are some free wavelengths. Thus there are transitions due to deflection only when the states are of the form (w, j). There are a symmetric set of transitions from port 2 to 1, when the states are of the type i, w. Combining all the steady-state equations for this case along with the condition that $\sum_{i=0}^{w} \sum_{j=0}^{w} \pi_{i,j} = 1$, the steady-state probabilities for the system $\pi_{i,j}$, $(0 \le i, j \le w)$ can be computed. The BLP for this system is obtained by considering the lossy states w, w and when the system is in state w, j or i, w but there is no deflection (with a probability $1 - p$). The BLP for ports 1 and 2 can be obtained as

$$PB_1 = \pi_{w,w} + (1 - p) \sum_{k=0}^{w-1} \pi_{w,k}, \qquad (4.81)$$

$$PB_2 = \pi_{w,w} + (1 - p) \sum_{l=0}^{w-1} \pi_{l,w}. \qquad (4.82)$$

If the probability of deflection is different at each port, say p_1 and p_2, the transition probabilities and the steady-state equations can be modified appropriately to include the effect of them.

4.4.2.4 Accurate Model Without Wavelength Conversion

For a node without wavelength conversion, the two-dimensional Markov chain that describes the system is given in Fig. 4.13.

Since each wavelength is modeled separately, the state of the system (i, j) indicates the availability of the wavelength on each port. Similar to the approximate model, 1 indicates that the wavelength is occupied so that an arriving burst is deflected with a probability p and 0 indicates that the wavelength is free. Therefore, when the system is in states $(1, 0)$ and $(0, 1)$, a burst is deflected with a probability p. If the burst is not deflected, it is lost. Hence, along with a possibility of the burst not being deflected in these states, bursts are also lost if the system is in state $(1, 1)$. BLP at port 1 and port 2 is thus given by

$$PB_1 = \pi_{1,1} + (1 - p)\pi_{1,0}, \qquad (4.83)$$
$$PB_2 = \pi_{1,1} + (1 - p)\pi_{0,1}. \qquad (4.84)$$

For any number of ports without wavelength conversion, the set of linear equations similar to the above can be simply solved to obtain the loss probabilities.

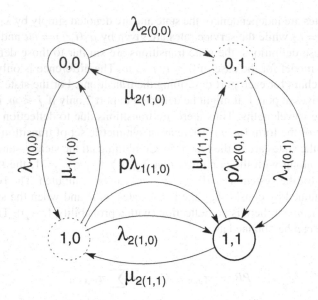

Fig. 4.13 State transitions in the accurate model for a node without wavelength conversion

In general, for a node with N ports, the system can be modeled by an N-dimensional Markov chain but the complexity of solving the system of equations is higher as N increases. Therefore, if the load is reasonable, BLP can be obtained easily by modeling the system using the approximate model given earlier.

4.4.3 Stabilizing Deflection Routing

It is well known that deflection routing may destabilize circuit switching as well as packet switching networks [18, 23]. It was observed that deflection routing causes instability in the network operation leading to a sudden increase in the blocking and a reduction in the carried load beyond a particular offered load [1, 26]. This occurs due to the fact that the routes used for transmitting bursts for the first time and that for the deflected bursts would have had some common links. A deflection from the first blocking thus increases the load along some other route causing further deflections. The effect of this increased load causes instability which is similar to the congestion collapse in the Internet. It is interesting to see if the same kind of instability is caused in OBS networks which is expected by intuition. Since the increased load due to the deflection routing might result in higher contention losses, oscillations in BLP are possible. This section demonstrates that instability is caused due to deflection routing even in the case of OBS networks [52]. There are a few proposals to stabilize OBS networks in the presence of deflection routing which are also discussed here [51, 53].

The primary reason for the instability due to deflection routing is that the routes used by the primary and the deflected bursts have some common links. For each source–destination pair, the primary route is an ordered set of links between the source and the destination nodes used for reservation. A deflection route on the other hand is an ordered set of links between any node along the primary route (or the source node) to the destination node. For deflection routing to be effective, the first link of each deflection route must be distinct from the first link of any other deflection route or a primary route [51]. A simple deflection routing protocol would have to check if the primary route is available for a burst arriving at an OXC failing which the burst is sent along one of the deflection routes. The deflection routing protocol can limit the number of deflection routes computed to satisfy the link disjointness property mentioned earlier. To limit the number of deflections, some protocols that deflect a burst multiple times [47] or deflect a burst using state-dependent parameters [13] may also be used. But for the purpose of simplifying the analysis, it is assumed here that a burst can be deflected only once along any one of the deflection routes [51]. It is also observed through simulations that the benefit obtained by allowing multiple deflections per burst is not significant compared to allowing only single deflection along a route [52].

Assume that the bursts along the primary route arrive following a Poisson process and have exponentially distributed lengths. For simplicity it may also be assumed that the deflected bursts are generated according to an independent Poisson process so that the blocking probability at a node with full wavelength conversion is same across the network given by the Erlang B formula in Eq. 3.1. Let ρ_e denote the external load due to the bursts generated between a source–destination pair and ρ be the total load offered on a link due to the bursts between all the source–destination pairs passing through the link. The carried load on a link is the load due to the bursts that have not been blocked which includes the bursts arriving for the first time as well those that are deflected. The carried load can be written as

$$(1-p)\rho =$$
$$(1-p)\rho_e + p(1-p)\rho_e + p(1-p)^2\rho_e + p(1-p)^3\rho_e + \ldots + p(1-p)^{H-1}\rho_e,$$
$$(4.85)$$

where p is the blocking probability on a link computed using the Erlang B formula and H is the path length. Note that computation of the carried load in OBS networks is not exactly same as that in circuit-switched networks [18] due to the reduction in the load at each link along the path due to the bursts that are blocked. In circuit-switched networks the carried load would be simply given by

$$(1-p)\rho = (1-p)\rho_e + (H-1)(1-p)^{H-1}\rho_e$$

since the load carried along each link would be same. By assuming that $H = 4$ and rearranging Eq. 4.85, the external load can be written as

$$\rho_e = \frac{\rho}{1 + 3p - 3p^2 + p^3}.$$
$$(4.86)$$

Assuming that the blocking probability on each link can be assumed to be independent of that on the other link, the blocking probability on a path between a source–destination pair (BLP on a path) is given by [52]

$$PB = 3p^2 - 3p^3 + p^4. \tag{4.87}$$

To confirm that the deflection routing indeed leads to an instability in the OBS networks, the carried portion of the external load (given by $(\rho_e(1 - PB))$ and the BLP are plotted against the external load (ρ_e) [51]. Results show that the deflection routing in OBS networks indeed leads to an instability in the network operation just as observed in the case of circuit-switched networks. The unstable mode of operation is seen for a small range of load values. Trunk reservation is proposed as a method to stabilize the operation of the circuit-switched network in the presence of deflection routing [26]. Motivated by such a proposal, there are a few proposals to stabilize OBS networks with deflection routing.

4.4.3.1 Wavelength Reservation

In this method, the number of wavelengths that can be used by the deflected bursts is kept constant by reserving a set of wavelengths exclusively for the primary bursts. That is, a link does not admit deflected bursts if k out of w wavelengths are busy. To analyze the impact of wavelength reservation, let ρ_d be the load due to deflected bursts on a link which can be written as

$$\rho_d = \rho - \rho_e, \tag{4.88}$$

where ρ and ρ_e are as defined earlier. Due to wavelength reservation, the primary route which is link disjoint with the other routes consists of only one link so that the total offered load on a link is the sum of external load and the load due to deflected bursts. Let $p_i, 0 \leq i \leq w$, denote the probability that i wavelengths are busy in a link which is computed using

$$p_i = \begin{cases} \frac{\rho^i \cdot p_0}{i!}, & 1 \leq i \leq k \\ \frac{\rho_e^{i-k} \cdot \rho^k \cdot p_0}{i!}, & k < i \leq w \end{cases}, \tag{4.89}$$

where the normalization factor p_0 is determined by equation $\sum_{i=0}^{w} p_i = 1$ and $w - k$ wavelengths are reserved exclusively for the primary bursts. The blocking probability for the primary bursts is simply given by p_w which is

$$P_p = p_w = \frac{\rho_e^{w-k} \rho^k p_0}{w!} \tag{4.90}$$

because a primary burst is blocked if all the wavelengths on a link are busy. Since the deflected bursts are blocked if more than k wavelengths are busy, the blocking probability of deflected bursts is given by

$$P_d = \sum_{i=k}^{w} \frac{\rho_e^{i-k} \rho^k p_0}{i!}. \tag{4.91}$$

For a link on a path with four hops, adding the total carried load on a link due to primary bursts and deflected bursts gives

$$(1 - P_p)\rho = (1 - P_p)\rho_e + P_p(1 - P_d)\rho_e + P_p(1 - P_d)^2\rho_e + P_p(1 - P_d)^3\rho_e \tag{4.92}$$

so that ρ_e can be written as

$$\rho_e = \frac{\rho}{1 + 3P_p - 3P_p P_d + P_p P_d^2}. \tag{4.93}$$

Assuming that the blocking in each link is independent, the path blocking probability can be written as

$$PB = 3P_p P_d - 3P_p P_d^2 + P_p P_d^3. \tag{4.94}$$

It can easily be verified by substituting $P_p = P_d$ that the above equation leads to Eq. 4.87. By plotting the carried portion of the external load and the path blocking probability against ρ, it can be seen that wavelength reservation stabilizes the network with deflection routing.

There is no rigorous criteria to decide the choice of k in this method [22]. If the value of k is chosen too small, then the stable behavior is not observed, whereas if k is taken too large, the benefit is not significant and there might be an onset of instability in the operation. Based on the simulation study in [52] it was observed that for $w = 120$, a value of k in the range of $[100, 110]$ would be a good choice to bring stability with deflection routing.

4.4.3.2 Preemptive Priority

Another method that helps in the stability of OBS network with deflection routing is to give the primary bursts a higher priority than the deflected bursts in scheduling the wavelengths [8]. With this approach, a primary burst tries to reserve a wavelength but if it is blocked a deflected burst that has already been scheduled can be preempted. This reduces the load due to the deflected bursts and maintains the stability of operation. The analysis presented earlier for the method of wavelength reservation holds even in this case except that P_p and P_d are recomputed using the following equations:

$$P_p = \frac{(\rho - \rho_d)^w/w!}{\sum_{i=0}^{w}(\rho - \rho_d)^i/w^i} \tag{4.95}$$

and

$$P_d = \frac{\rho \frac{\rho^w/w!}{\sum_{i=0}^{w} \rho^i/w^i} - (\rho - \rho_d)\frac{(\rho-\rho_d)^w/w!}{\sum_{i=0}^{w}(\rho-\rho_d)^i/w^i}}{\rho_d}. \tag{4.96}$$

Since the primary bursts have a higher priority over the deflected bursts, the blocking perceived by the primary bursts is a result of the load only due to the primary bursts. The blocking probability in such a case is obtained by using the load after subtracting ρ_d from the total load ρ. Similarly, for the blocking perceived by the deflected bursts, the numerator in the equation above is the blocked load of the deflected bursts on a link while the denominator is the load offered due to the deflected bursts. Computing the carried portion of the external load and path blocking probability is similar to the case of wavelength reservation with modified P_p and P_d.

By plotting the carried portion of the external load and the blocking probability against the external load for this method, it can be seen that there is no instability due to deflection routing. It can also be seen that the preemptive priority method yields a lower blocking probability compared to the wavelength reservation method [52]. The advantage of preemptive priority over the wavelength reservation method can be explained by observing that stability is guaranteed by preemption of the deflected bursts by the primary bursts. Preemption ensures that the blocking performance is no worse than the case where bursts are not deflected but simply blocked. Since the deflected bursts cannot influence the blocking performance of the primary bursts, the preemption events are more evident with increasing load due to deflected bursts, making deflection routing ineffective. The preemption method can be used in tandem also with the burst segmentation thus leaving at least a portion of the deflected burst for further transmission and thus salvaging some packets. In that case, the load due to the deflected bursts can be recomputed with the analysis used for the burst segmentation. Although, the packet loss probability is improved by using segmentation of the deflected bursts in case of contention with the primary bursts, the complexity associated with segmentation makes it unfavorable for use with deflection routing which already needs proper dimensioning of the offset time.

4.4.3.3 Analysis for Large Networks

The analysis for deflection routing given in the above sections is used to compute the blocking probability for a small symmetric network (a path of length 4). For a larger and asymmetric network, it is not easy to use such analysis based on a single variable. But in general, it is such larger asymmetric networks that need to be analyzed. The reduced-load Erlang fixed point analysis explained in Section 3.8 can be used for analyzing the BLP of large networks. In this section, the reduced-load Erlang fixed point method is used to analyze a general asymmetric JET OBS network with deflection routing and wavelength reservation although it can be used for the preemption method also [51, 52]. Assuming that the blocking in each link evolves independently, the reduced load due to the bursts that are blocked at the previous hops on a path is considered while computing the blocking on a link. The

overall BLP along a path is obtained by solving a set of fixed point equations itera-
tively.

Let the set of links $\mathscr{P}$ and $\mathscr{D}$ denote the set of primary and deflection routes. A
link $(r, s) \in \mathscr{P}$ if a primary burst traverses through (r, s) and likewise for deflection
route. Using the simple deflection policy proposed in this section, $\mathscr{P} \cap \mathscr{D} = 0$.
To index the links, suppose (r, s_1) and (r, s_2), $(s_1 \neq s_2)$ are two links along the
primary route and if $s_1 < s_2$, a BHP reserves the wavelength on the link (r, s_1)
before link (r, s_2). Let ρ_e be the external load offered on a link by the bursts between
a source–destination pair, $\rho_p^{(r,s)}$ be the load due to primary bursts on a link (r, s), and
$\rho_d^{(r,s)}$ be the load due to the deflected bursts on a link (r, s). Let $P_p^{(r,s)}$ be the blocking
probability of primary bursts on the link (r, s) and $P_d^{(r,s)}$ be the blocking for the
deflected bursts on the same link. If wavelength reservation is not used for stability,
then $P_d = P_p$ for all links (r, s) but with wavelength reservation $P_p^{(r,s)} < P_d^{(r,s)}$.
For simplicity we can use same variable to indicate the blocking probability and the
load for either a primary or a deflected burst on a link (r, s) depending on the route
it belongs to. So

$$p^{(r,s)} = \begin{cases} P_p^{(r,s)} & \text{if } (r, s) \in \mathscr{P} \\ P_d^{(r,s)} & \text{if } (r, s) \in \mathscr{D} \end{cases} \tag{4.97}$$

and

$$\rho^{(r,s)} = \begin{cases} \rho_p^{(r,s)} & \text{if } (r, s) \in \mathscr{P} \\ \rho_d^{(r,s)} & \text{if } (r, s) \in \mathscr{D}. \end{cases} \tag{4.98}$$

Using the independence property of the load and the blocking probability of each
link, the reduced load on a link can be written in a recursive form as

$$\rho^{(r,s)} = \rho^{(q,r)}(1 - p^{(q,r)}). \prod_{(i,j) \in S(r,s)} p^{(i,j)}, \tag{4.99}$$

where (q, r) is the link preceding (r, s) and $S(r, s)$ is the set of all links $\{(i, j) :
i = r, j < s\}$. The above equation can be solved to obtain $\rho^{(r,s)}$ for all r except the
source node x. For $r = x$, the load is given by the recursion

$$\rho^{(x,s)} = \rho_e \prod_{(i,j) \in S(x,s)} p^{(i,j)}. \tag{4.100}$$

The recursive equation given above is solved for each source–destination pair
and for each (r, s) the resulting $\rho^{(r,s)}$ is obtained which are then added to obtain the
total load offered by both the primary and the deflected bursts on a link due to all
source–destination pairs.

As a part of the wavelength reservation policy for stability, let $w^{(r,s)}$ be the total
number of wavelengths in a link (r, s) of which $w^{(r,s)} - k^{(r,s)}$ are reserved for primary

bursts. By modeling each link as an $M/M/1/w^{(r,s)}$ queue, let $p_j^{(r,s)}$ be the probability that $1 \leq j \leq k^{(r,s)}$ out of $w^{(r,s)}$ wavelengths are used on the link (r, s) which can be written as

$$p_j^{(r,s)} = (\rho_p^{(r,s)} + \rho_p^{(r,s)}) \cdot p_0^{(r,s)} / j! \qquad (4.101)$$

and for $k^{(r,s)} < j \leq w^{(r,s)}$,

$$p_j^{(r,s)} = \rho_p^{j-k^{(r,s)}} (\rho_p^{(r,s)} + \rho_p^{(r,s)}) \cdot p_0^{(r,s)} / j!, \qquad (4.102)$$

where $p_0^{(r,s)}$ is determined using the condition $\sum_{j=0}^{w^{(r,s)}} p_j^{(r,s)} = 1$. The iterative equations 4.101 and 4.102 can be solved to obtain $p_j^{(r,s)}$ for each link (r, s).

To obtain the blocking probabilities of primary and deflected bursts on a link (r, s), note that the primary bursts are lost if all the $w^{(r,s)}$ wavelengths are busy which occurs with a probability

$$P_p^{(r,s)} = p_{w^{(r,s)}}^{(r,s)}. \qquad (4.103)$$

Similarly, a deflected burst is dropped if $k^{(r,s)}$ or more wavelengths are busy which occurs with a probability of

$$P_d^{(r,s)} = \sum_{j=k^{(r,s)}}^{w^{(r,s)}} \frac{\rho_p^{(r,s)j-k^{(r,s)}} \cdot p_0^{(r,s)} \cdot (\rho_p^{(r,s)}) + \rho_d^{(r,s)})^{w^{(r,s)}}}{j!}. \qquad (4.104)$$

Observe that Eqs. 4.99, 4.100, 4.103, and 4.104 are coupled because $\rho_p^{(r,s)}$ and $\rho_d^{(r,s)}$ depend on the link blocking probabilities $P_p^{(r,s)}$ and $P_d^{(r,s)}$ and vice versa. A unique solution for all these equations solved simultaneously is known as EFP which represents the stationary link blocking probabilities denoted by $P_p^{*(r,s)}$ and $P_d^{*(r,s)}$. Although the existence of an EFP is not guaranteed, the successive substitution algorithm provided in [51] is an efficient method to find one.

In the successive substitution method, the link loads are initialized to some random values and for each link the blocking probabilities are updated according to Eqs. 4.103 and 4.104. This procedure is carried out till the blocking probabilities do not differ significantly between any two iterations. The final values of the blocking probability are defined as the stationary values which are used to recompute the load due to the primary and the deflected bursts using Eqs. 4.99 and 4.100. The load thus computed is used in the iterative procedure to compute the blocking probability till the values converge.

Once an EFP is obtained, to compute the end-to-end blocking probability of the path for a source–destination pair, let PB_i denote the blocking probability at a node i before reserving the wavelength on an outgoing link at i. Let the stationary blocking probabilities be defined for the primary and deflected bursts similar to Eq. 4.97 by a common variable $P^{*(r,s)}$. With the assumption that the blocking is independent in each link,

$$PB_i = \sum_{j \in R(i)} P_j (1 - P^{*(i,j)}) . \prod_{(i,j) \in S(i,j)} P^{*(i,j)} + P^{*(r,\max R(i))} . \prod_{(i,j) \in S(i,\max R(i))} P^{*(i,j)},$$
(4.105)

where $R(i) = \{s : (r,s) \in \mathscr{P} \text{ or}(r,s) \in \mathscr{D}\}$. The recursion can be solved to obtain PB_i for each node i, which is not a destination node. For the destination node, say y, $P_y = 0$.

Finally, for all the source–destination pairs indexed by $t = 1, 2, \ldots n$, the overall blocking probability for the network can be obtained by

$$BLP = \frac{1}{\sum_{t=1}^{n} \rho_e^t} \sum_{t=1}^{n} \rho_e^t PB_s^t,$$
(4.106)

where PB_s is the BLP for a source–destination pair s. The overall burst load carried in the network is given by

$$L = \sum_{t=1}^{n} (1 - PB_s^t) . \rho_e^t.$$
(4.107)

Thus, the BLP and the load in any general asymmetric JET-based OBS network using deflection routing can be evaluated.

4.5 Analysis of Fiber Delay Lines

FDLs can be used to buffer optical bursts for a short duration so that they help to resolve contention among multiple bursts [11]. Particularly, in OBS networks due to the formation of several voids in the schedule, FDLs can avoid burst losses significantly by delaying the burst to fit into a void. Since the other forms of optical buffers are immature for deployment and expensive, FDLs are used as a technology for optical buffer in most of the cases [33]. Although the architecture of an OBS node does not mandate the use of optical buffers, the BLP can be significantly reduced by using FDL-based buffers [49].

Fundamentally FDLs consist of a long length of fiber that delays the optical burst by additional propagation through the fiber. Though the delay provided by the FDLs is very short (few kilometers of fiber can provide delay in the order of microseconds), it has been shown through analysis and simulations that there is significant performance gain in BLP due to the usage of FDLs as a contention resolution technique [16, 50]. In association with full wavelength conversion, which resolves contentions by wavelength dimensioning, FDLs can reduce the BLP significantly. The basic configuration of an optical switch with output buffering is discussed in detail in [40, 50]. Optical buffers based on the FDLs can be designed in many ways of which there are fixed delay buffers [49, 50] and variable delay buffers [21, 40, 50]. In combination with wavelength conversion, the physical FDL buffers effectively can provide larger number of buffers termed as virtual buffers. If there are F physical buffers with w wavelengths among which complete conversion is possible, the number of virtual buffers is wF. Both fixed and variable delay buffers are equally good in reducing the BLP in an event of contention.

4.5.1 Exact Markovian Model

Considering the benefit of FDLs in reducing contention losses, it would be interesting to analyze the performance of OBS networks with FDLs at the core nodes. Analytical modeling of the OBS node with FDLs is challenging due to the difference between these and the conventional electronic buffers. In conventional electronic buffers, a packet gets queued if there is sufficient space and remains there till bandwidth is available. However, in FDLs the delay provided is limited by the maximum propagation time through the length of the fiber. In effect the FDL can be considered as a buffer with a bound on the delay. Apart from that a burst is stored in the FDL only if the additional delay required for the service is less than the maximum delay that can be provided by the FDL. So whenever the time for which a burst needs to be delayed before it is serviced exceeds the maximum delay provided by the FDL, it is dropped before being queued. This phenomenon is known as *balking* in queueing theory which is a special case of a queue with impatient customers or queue with discouraged arrivals [19]. In the previous chapter, models for an OBS node with FDL used an $M/M/W/D$ queue where the maximum number of bursts that can be stored in the queue (FDL) is limited to D which depends on the number of virtual FDLs. In this section we discuss the models for OBS node with FDLs that include the effect of finite delay and balking.

A simple model was proposed for an FDL with one wavelength per fiber in [7] that investigates the impact of delay time of each unit in degenerate FDLs. Though the model in [7] uses a queue with balking it is restricted only to a delay in integral units and is not generic for use in OBS networks. The analysis discussed here is generic for an OBS node with FDLs and works well for any length of FDLs, distribution of burst length, and inter-arrival time. At an OBS node with FDLs, apart from the wavelength reservation module, the scheduling algorithm also has an FDL reservation module. When a new burst arrives, if there is no free wavelength at that time or for the entire duration of the burst, the scheduling algorithm computes the minimum waiting time after which the burst can be serviced. If the minimum waiting time is less than the maximum delay that can be provided by the FDL, the FDL is reserved and the wavelength is reserved after the delay. The minimum waiting time is simply the time for which all the wavelengths are busy, i.e., the sum of the residual time of service for the bursts being currently serviced and the durations of all the bursts that have already been scheduled during the burst duration. Consider an OBS node with F variable delay FDL buffers and full wavelength conversion among w

wavelengths. Each FDL on a wavelength is capable of providing a delay between zero and D delay units. If a burst is directed to an FDL with a delay requirement less than D units, it is buffered. In this way wF bursts can be simultaneously serviced by the FDL.

With the above architecture, an OBS node with FDLs can be modeled using a multi-dimensional CTMC to evaluate the blocking probability of a burst [30, 31]. Assuming that a new burst chooses an output port with uniform probability, it is enough to model the behavior of a typical output port of a switch. Assume that the bursts arrive following a Poisson distribution with a mean rate of λ bursts/second and the burst duration is exponentially distributed with a mean $1/\mu$ seconds. At a given time t, let the state of the system be characterized by the tuple $(s_1, s_2, \ldots, s_w, v)$, where s_i is the number of bursts assigned to a wavelength i and $0 \le v \le wF$ be the number of virtual buffers that are busy at time t. Among the data bursts assigned to a wavelength i, only one burst is serviced while the others are sent to the FDL unit. Without the loss of generality, consider that $w = 2$, so that the state of the system at a given time is represented by (s_1, s_2, v). According to the values of s_1 and s_2, four cases are possible to describe the system as given below:

- $s_1 = 0$, $s_2 = 0$: Since both the wavelengths are idle, there is no burst loss in this state. A new burst arrival leads to the next state which is either $(1, 0, v)$ or $(0, 1, v)$, with rate $\lambda/2$ assuming that both the transitions are equi-probable. The other possible transition from this state is $(0, 0, v - 1)$ with rate $v\mu$ if $v > 0$.
- $s_1 = 0$, $s_2 > 0$: There is no blocking in this state because one of the wavelengths is free. Since the first wavelength is free, a new burst is assigned to it leading to the state $(1, s_2, v)$ with probability λ. If a burst is serviced, the next state is either $(0, s_2, v - 1)$ or $(0, s_2 - 1, v)$ for $v > 0$ with rate $v\mu$ and μ, respectively.
- $s_1 > 0$, $s_2 = 0$: This state is similar to the above case and has similar transitions.
- $s_1 > 0$, $s_2 > 0$: In this case, the new data bursts are blocked in the wavelength reservation phase and moved to the FDLs. A new burst arrival changes the state to either $(s_1 + 1, s_2, v + 1)$ or $(s_1, s_2 + 1, v + 1)$ with a probability λ depending on which wavelength is assigned. The departure of a burst causes a transition to $(s_1 - 1, s_2, v)$ or $(s_1, s_2 - 1, v)$ with a rate μ, and $(s_1, s_2, v - 1)$ with a rate $v\mu$ for $v > 0$.

At any time t, in state (s_1, s_2, v), let the duration for which a wavelength i would be busy be D_i. So any incoming burst is assigned to wavelength j, if $D_j < D_i$ for all i and $D_j < D$. For the case of two wavelengths, a burst arrival leads to the state $(s_1 + 1, s_2, v)$ if $D_1 < D$ and $D_1 < D_2$, and $v < wF$. Let P_i be the probability that a new burst is discarded if i bursts are already scheduled and $M_{i,j}$ be the probability that $D_1 < D_2$, given that the number of bursts already scheduled on wavelengths w_1 and w_2 are i and j, respectively. The probability of transition to a state $(s_1 + 1, s_2, v)$ when $v < wF$ can be written as

$$\lambda[Pr\{D_1 < D_2 < D\} + Pr\{D_1 < D < D_2\}] = \lambda[(1 - P_{s_1})(1 - P_{s_2})M_{s_1,s_2} \\ + (1 - P_{s_1})P_{s_2}]. \quad (4.108)$$

Similarly, the probability of transition to a state $(s_1, s_2 + 1, v)$ with $v < wF$ is given by

$$\lambda[Pr\{D_2 < D_1 < D\} + Pr\{D_2 < D < D_1\}] = \lambda[(1 - P_{s_2})(1 - P_{s_1})M_{s_2,s_1} + (1 - P_{s_2})P_{s_1}]. \quad (4.109)$$

To evaluate P_i, let the residual service time of the data burst currently under service be R and suppose that there are $i - 1$ bursts still to be serviced (already scheduled before t). Let the service time of the bursts which is assumed to be a sequence of exponentially distributed independent random variables with mean $1/\mu$ be denoted by $\{X_i\}$. As mentioned earlier, a new burst is blocked (not even assigned to FDL) if the total delay (total residual service time of all the bursts) exceeds the maximum delay that can be provided by the FDL. Therefore, the blocking probability of a burst arriving at t is given by

$$P_i = Pr\left\{R + \sum_{k=1}^{i-1} X_k > D\right\}. \quad (4.110)$$

Using the memoryless property of exponentially distributed random variables,

$$P_i = Pr\left\{\sum_{k=0}^{i-1} X_k > D\right\} \quad (4.111)$$

$$= e^{-\mu D} \sum_{j=0}^{i-1} \frac{(\mu D)^j}{j!}. \quad (4.112)$$

Note that the blocking probability as seen from Eq. 4.111 is a complementary distribution of an i-stage Erlangian random variable where i is the number of bursts already scheduled at time t.

To obtain $M_{i,j}$, let the service times of bursts on wavelengths w_1 and w_2, which form a sequence of exponentially distributed random variables, be denoted by $\{X_l\}_{1 \leq l \leq i}$ and $\{Y_l\}_{1 \leq l \leq j}$, respectively. By definition of $M_{i,j}$, for any given number of bursts $i, j \geq 0$ already scheduled,

$$M_{i,j} = Pr\left\{\sum_{l=1}^{i} X_l < \sum_{l=1}^{i} Y_l\right\} \quad (4.113)$$

$$= \sum_{l=0}^{j-1} \binom{l+i-1}{l} \frac{1}{2^{l+i}}. \quad (4.114)$$

Substituting P_i and $M_{i,j}$ into Eq. 4.108 gives the probability of a state transition when a new burst arrives. Let π_{ijk} denote the steady-state probability of the system being in state (i, j, k) and the BLP in this state be denoted by PB_{ijk}. From the analysis so far, the BLP in a state (i, j, k) is given by

$$PB_{ijk} = P_i P_j + (1 - P_i P_j)\delta_{kr}, \qquad (4.115)$$

where $r = wF$ is the number of virtual FDLs and $\delta_{kr} = 0$ if $k \neq r$ and 1 otherwise.
The overall BLP can be obtained by considering the BLP over all possible states as

$$BLP = \sum_{i=0}^{\infty} \sum_{j=0}^{\infty} \sum_{k=0}^{r} \pi_{ijk} PB_{ijk}. \qquad (4.116)$$

The Markov chain model described above can be solved using the state-truncation
techniques [25] which compute the BLP to a desired level of accuracy for a given
state space. The model provides insights into the performance of OBS networks
with FDLs and gives an easy way of computing BLP as a function of the number
of wavelengths, load, and the maximum delay given by FDL. This model shows
the difference in modeling FDL-based buffers and modeling FDLs as conventional
electronic buffers. However, the analytical model becomes increasingly complex
with increasing number of wavelengths because the dimension of the state space
grows with w.

4.5.2 Asymptotic Approximation Models

To make the exact model derived in the previous section tractable for large values
of w, this section discusses a queueing model with balking for short FDLs and a
different queueing model for long FDLs. The exact model derived above does not
explicitly consider the length of the burst compared to the length of the FDL. In
other words, it evaluates the probability of a burst not getting an FDL and thereby
getting blocked. However, if D is either very small or very large, simpler models
can be obtained by some approximations which are useful to evaluate BLP for even
large values of w. To model both short and long FDL regimes, observe that a burst
can be blocked in two phases: wavelength reservation phase and FDL reservation
phase. Let E_W denote the event that a burst is blocked in the wavelength reservation
phase and E_F be the event that a burst is blocked in the FDL reservation phase. The
BLP in the wavelength reservation phase is thus $PB_W = Pr\{E_W, T > D\}$, where T
is the duration for which the burst has to wait before being serviced. Similarly, the
BLP in the FDL reservation phase is $PB_F = Pr\{E_W, T \leq D, E_F\}$. The total BLP
is the sum of the BLP during the two phases. For short FDLs, i.e., $D \approx 0$, if bursts
are blocked in the wavelength reservation phase, it is unlikely that FDL is capable
of providing required delay. So, most of the bursts are blocked in the wavelength
reservation phase itself, i.e., $PB_F \ll PB_W$ and $BLP \approx PB_W$. Conversely, for long
FDLs, i.e., $B \to \infty$, almost all the bursts that require FDL are serviced later due to
the availability of sufficiently long FDLs. Bursts blocked in this case are only those
that find no FDLs or find all the wavelengths busy. In this case $PB_W \ll PB_F$ and

$BLP \approx PB_F$. In the regime of short FDLs, an accurate model must consider the balking property of the queue, whereas for long FDLs the maximum delay provided by FDL, D must be characterized.

4.5.2.1 Balking Model for Short FDLs

The focus in this regime is mainly on the BLP in the wavelength reservation phase where it is assumed that the FDL is reserved only if the waiting time is less than D. Based on the theory of queues with balking [19] an OBS node with FDL can be modeled as an $M/M/w$ queue with balking [29]. Let the state of the system be described by the number of bursts in the system, which is denoted by the stochastic process $\{X_t, t > 0\}$. The system thus described is a continuous-time discrete Markov chain as illustrated in Fig. 4.14.

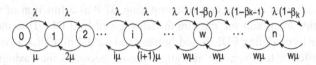

Fig. 4.14 Continuous-time discrete Markov chain representation of an OBS node with FDL

Let λ_i and $\mu_i = \min\{i, w\}\mu$ be the arrival rate and the service rate of bursts when the system is in state i, i.e., there are i bursts already scheduled. A burst is never lost if $i < w$ since it always finds a suitable wavelength so that $\lambda_i = \lambda$. However, if $i \geq w$, a burst is lost if there is no free wavelength till $t + D$ or if the number of bursts serviced during D is less than $i - w + 1$. For $i \geq w$, the probability that a burst is not lost is therefore equal to the probability that at least $i - w + 1$ bursts are serviced during D time units. Thus the BLP in state i is given by

$$PB_i = 1 - \int_0^D e^{-w\mu x} \frac{w\mu(w\mu x)^{i-w}}{(i-w)!} dx$$

$$= 1 - \sum_{k=0}^{i-w} e^{-w\mu D} \frac{(w\mu D)^k}{k!}. \tag{4.117}$$

The steady-state probability of state $i \geq 0$, denoted by π_i, is derived using λ_i and μ_i as

$$\pi_0 = \left[1 + \sum_{k=0}^{\infty} \prod_{i=0}^{k-1} \frac{\lambda_i}{\mu_{i+1}} \right]^{-1}, \tag{4.118}$$

$$\pi_i = \pi_0 \prod_{j=0}^{i-1} \frac{\lambda_j}{\mu_{j+1}}, \quad i \geq 1 \tag{4.119}$$

from which the overall BLP for short FDLs is obtained as

$$BLP = \sum_{i=w}^{\infty} \pi_i PB_i. \tag{4.120}$$

4.5.2.2 Queueing Model for Long FDLs

For long FDLs, the delay that can be provided by the FDL is large, i.e, $D \gg 0$ so that a burst is mostly blocked due to the unavailability of FDLs rather than the waiting time exceeding D. In the asymptotic limit of $D \to \infty$, each virtual buffer can provide any amount of delay requested by the burst after the wavelength reservation phase. In this regime, the FDL can be modeled as an $M/M/w/w + wF$ queue where the number of virtual buffers is wF [30]. The BLP in the regime of long FDLs is thus given by

$$BLP = \frac{\frac{\rho^{w+wF}}{(w^{wF}.w!)}}{\sum_{k=0}^{j-1} \rho^k/k! + \sum_{k=w}^{w+wF} \rho^k/(w^{k-w}.w!)}. \tag{4.121}$$

This is the same model proposed in [43, 50] described in Section 3.2.

Although, variable delay FDLs are used in the analysis, the delay can be provided between 0 and D in units of a fundamental delay value d. The effect of delay granularity was studied in [7] by applying a balking queueing model to evaluate the BLP. Suppose that a burst arrives at t to find that there is a waiting time of t_w till an output wavelength is free. Ideally, the data burst is delayed by a value of t_w but, in practice due to the granularity of the delay value, the actual delay would be $d_a = \lceil t_w/d \rceil d$. Due to this the wavelength would remain idle for a duration of $t_l = d_a - t_w$. The idle time is modeled as a virtual increase in the burst length which is uniformly distributed in the interval $[0, d]$ due to the assumption that the burst lengths are independently distributed. Let $\overline{L}$ be the mean length of the burst and $\overline{L_e}$ be the mean increase in the burst length related to $\overline{L}$ by

$$\overline{L_e} = (1 - p)\overline{L} + p(\overline{L} + d/2) = \overline{L} + pd/2, \tag{4.122}$$

where p is the probability that an incoming burst is delayed. The load to the output port is then given by $\rho = \lambda \overline{L}$ and the effective load due to the increase in the burst length is $\rho_e = \lambda \overline{L_e}$. The distribution of the increased burst length $\overline{L_e}$ cannot be taken to be exponential, but for tractability it is assumed to be so. Using an iterative procedure, $\overline{L_e}$ can be obtained which is used in the analysis either in the exact or approximate models to obtain the BLP considering the FDL granularity. Numerical and simulation results in [31] show that FDLs can provide significant performance gain in OBS networks by reducing the BLP by two to three orders of magnitude.

Though the model for an OBS node with FDL discusses both short and long FDL regimes, numerical results in [29–31] demonstrate the accuracy of the model only for a small number of wavelengths. For tractability of the model, the system with only two wavelengths is considered and it was also pointed out in the discussion that the complexity is high for larger number of wavelengths. For this sake, the approximate models that neglect blocking in either one of the two phases, wavelength reservation and FDL reservation, are provided. However, in [34] it was shown through the simulations that the analysis given above is not accurate for larger number of wavelengths. Accuracy of the model is particularly low in the cases when the BLP is low or the number of wavelengths is large. The model is also inaccurate when the burst service times are close to the length of the FDL. The inaccuracy of the model can be explained from the way the probability of balking is defined. In any given state, which is defined based on the number of bursts in the system, the probability that a burst is blocked because the waiting time is larger than the delay possible in the FDL depends only on the number of bursts in the system. But, in reality it depends also on the residual life of the system, which includes all the wavelengths and the FDLs, captured by the random variable representing the inter-departure times of the bursts (defined in Eq. 4.113). It would be interesting to explore better models for the OBS nodes with FDLs to improve accuracy of the analysis even for a larger number of wavelengths.

References

1. Akimpelu, J.M.: The overload performance of engineered networks with hierarchical and nonhierarchical routing. Bell Systems Technical Journal 63(7), 1261–1281 (1984)
2. Bannister, J., Borgonovo, F., Fratta, L., Gerla, M.: A performance model of deflection routing in multibuffer networks with nonuniform traffic. IEEE Transactions on Networking 3(5), 509–520 (1995)
3. Bertsekas, D., Gallager, R.: Data Networks. Prentice Hall, USA (1992)
4. Bononi, A., Forghieri, F., Prucnal, P.R.: Analysis of one-buffer deflection routing in ultra-fast optical mesh networks. In: Proceedings of IEEE INFOCOM, pp. 303–311 (1993)
5. Borgonovo, F., Fratta, L., Bannister, J.: Unslotted deflection routing in all-optical networks. In: Proceedings of IEEE GLOBECOM, pp. 119–125 (1993)
6. Borgonovo, F., Fratta, L., Bannister, J.: On the design of optical deflection-routing networks. In: Proceedings of IEEE INFOCOM, pp. 120–129 (1994)
7. Callegati, F.: Optical buffers for variable length packets. IEEE Communications Letters 4(9), 292–294 (2000)
8. Cameron, C., Zalesky, A., Zukerman, M.: Prioritized deflection routing in optical burst switching networks. IEICE Transactions on Communications E88-B(5), 1861–1867 (2005)
9. Chen, Y., Wu, H., Xu, D., Qiao, C.: Performance analysis of optical burst switched node with deflection routing. In: Proceedings of IEEE GLOBECOM, pp. 1355–1359 (2003)
10. Chich, T., Cohen, J., Fraigniaud, P.: Unslotted deflection routing: A practical and efficient protocol for multihop optical networks. IEEE Transactions on Networking 9(1), 47–59 (2001)
11. Chlamtac, I., et. al.: CORD: Contention resolution by delay lines. IEEE Journal on Selected Areas in Communications 14(5), 1014–1029 (1996)
12. Choudhury, A.K., Li, V.O.K.: An approximate analysis of the performance of deflection routing in regular networks. IEEE Journal on Selected Areas in Communications 11(8), 1302–1316 (1993)
13. Chung, S.P., Kashper, A., Ross, K.W.: Computing approximate blocking probabilities for large loss networks with state-dependent routing. IEEE Transactions on Networking 1(1), 105–115 (1993)
14. Detti, A., Eramo, V., Listanti, M.: Performance evaluation of a new technique for IP support in a WDM optical network: Optical composite burst switching (OCBS). Journal of Lightwave Technology 20(2), 154–165 (2002)

15. Du, Y., Zhu, C., Zheng, X., Guo, Y., Zhang, H.: A novel load balancing deflection routing strategy in optical burst switching networks. In: Proceedings of SPIE OFC/NFOEC, pp. 25–27 (2007)
16. Fan, P., Feng, C., Wang, Y., Ge, N.: Investigation of the time-offset-based QoS support with optical burst switching in WDM networks. In: Proceedings of IEEE ICC, pp. 2682–2686 (2002)
17. Gambini, P., et.al.: Transparent optical packet switching: Network architecture and demonstrators in the KEOPS project. IEEE Journal on Selected Areas in Communications 16(7), 1245–1259 (1998)
18. Girard, A.: Routing and Dimensioning in Circuit-Switched Networks. Addison Wesley, USA (1990)
19. Gross, D., Harris, C.M.: Fundamentals of Queueing Theory. John Wiley & Sons, USA (2003)
20. Hsu, C.F., Liu, T.L., Huang, N.F.: Performance analysis of deflection routing in optical burst-switched networks. In: Proceedings of IEEE INFOCOM, pp. 66–73 (2002)
21. Hunter, D., Chia, M., Andonovic, I.: Buffering in optical packet switches. Journal of Light-Wave Technology 16(12), 2081–2094 (1998)
22. Kaniyil, J., Hagiya, N., Shimamoto, S., Onozato, Y., Noguchi, S.: Structural stability aspects of alternate routing in non-hierarchical networks under reservation. In: Proceedings of IEEE GLOBECOM, pp. 797–801 (1992)
23. Kelly, F.P.: Blocking probabilities in large circuit switched networks. Advances in Applied Probability 18, 473–505 (1986)
24. Kim, S., Kim, N., Kang, M.: Contention resolution for optical burst switching networks using alternative routing. In: Proceedings of IEEE ICC, pp. 2786–2790 (2002)
25. Kleinrock, L.: Queueing Systems. Vol. 1, John Wiley and Sons, USA (1975)
26. Krupp, R.S.: Stabilization of alternate routing networks. In: Proceedings of IEEE ICC, pp. 31.2.1–31.2.5 (1982)
27. Lee, S., Kim, L., Song, J., Griffith, D., Sriram, K.: Dynamic deflection routing with virtual wavelength assignment in optical burst-switched networks. Photonic Network Communications 9(3), 347–356 (2005)
28. Lee, S., Sriram, K., Kim, H., Song, J.: Contention-based limited deflection routing protocol in optical burst-switched networks. IEEE Journal of Selected Areas in Communications 23(8), 1596–1610 (2005)
29. Lu, X., Mark, B.L.: Analytical modeling of optical burst switching with fiber delay lines. In: Proceedings of IEEE/ACM International Symposium on Modeling, Analysis Simulation of Computer, Telecommunication Systems, pp. 501–506 (2002)
30. Lu, X., Mark, B.L.: A new performance model of optical burst switching with fiber delay lines. In: Proceedings of IEEE ICC, pp. 1365–1369 (2003)
31. Lu, X., Mark, B.L.: Performance modeling of optical burst switching with fiber delay lines. IEEE Transactions on Communications 52(12), 2175–2183 (2004)
32. Magana, E., Morato, D., Izal, M., Aracil, J.: Evaluation of preemption probabilities in OBS networks with burst segmentation. In: Proceedings of IEEE ICC, pp. 1646–1650 (2005)
33. Maier, M., Reisslein, M.: Trends in optical switching techniques: A short survey. IEEE Network 22(6), 42–47 (2008)
34. Morato, D., Aracil, J.: On the use of balking for estimation of the blocking probability for OBS routers with FDL lines. In: Proceedings of IEEE ICOIN, pp. 399–408 (2006)
35. Neuts, M., Rosberg, Z., Vu, H.L., White, J., Zukerman, M.: Performance analysis of optical composite burst switching. IEEE Communications Letters 6(8), 346–348 (2002)
36. Neuts, M., Vu, H.L., Zukerman, M.: Insights into the benefit of burst segmentation in optical burst switching. In: Proceedings of Conference on Optical Internet and Photonics in Switching, pp. 126–128 (2002)
37. Phuritatkul, J., Ji, Y.: Buffer and bandwidth allocation algorithms for quality of service provisioning in WDM optical burst switching networks. In: Lecture notes in Computer Science, Vol. 3079, pp. 912–920 (2004)

38. Phuritatkul, J., Ji, Y., Yamada, S.: Proactive wavelength pre-emption for supporting absolute QoS in optical-burst-switched networks. Journal of Lightwave Technology **25**(5), 1130–1137 (2007)
39. Phuritatkul, J., Ji, Y., Zhang, Y.: Blocking probability of a preemption-based bandwith-allocation scheme for service differentiation in OBS networks. Journal of Lightwave Technology **24**(8), 2986–2993 (2006)
40. Ramaswami, R., Sivarajan, K.N.: Optical Networks: A Practical Perspective. Morgan Kauffmann, USA (2002)
41. Rosberg, Z., Vu, H.L., Zukerman, M.: Burst segmentation benefit in optical switching. IEEE Communications Letters **7**(3), 127–129 (2003)
42. Tan, C.W., Mohan, G., Li, J.C.S.: Achieving multi-class service differentiation in WDM optical burst switching networks: A probabilistic preemptive burst segmentation scheme. IEEE Journal on Selected Areas in Communications **24**(12), 106–119 (2006)
43. Turner, J.: Terabit burst switching. Journal of High Speed Networks **8**(1), 3–16 (1999)
44. Vokkarane, V.M., Jue, J.P.: Prioritized burst segmentation and composite burst-assembly techniques for QoS support in optical burst-switched networks. IEEE Journal on Selected Areas in Communications **21**(7), 1198–1209 (2003)
45. Vokkarane, V.M., Jue, J.P.: Introduction to Optical Burst Switching. Springer, USA (2004)
46. Vokkarane, V.M., Jue, J.P., Sitaraman, S.: Burst segmentation: An approach for reducing packet loss in optical burst switched networks. In: Proceedings of IEEE ICC, pp. 2673–2677 (2002)
47. Wang, X., Morikawa, H., Aoyama, T.: Burst optical deflection routing protocol for wavelength routing WDM networks. In: Proceedings of SPIE OptiComm, pp. 257–266 (2000)
48. Wang, X., Morikawa, H., Aoyama, T.: Burst optical deflection routing protocol for wavelength routing WDM networks. Optical Network Magazine **3**(6), 12–19 (2002)
49. Xoing, Y., Vandenhoute, M., Cankaya, C.: Control architecture in optical burst-switched WDM networks. IEEE Journal of Selected Areas in Communications **18**(10), 1838–1851 (2000)
50. Yoo, M., Qiao, C., Dixit, S.: QoS performance of optical burst switching in IP-over-WDM networks. IEEE Journal on Selected Areas in Communications **18**(10), 2062–2071 (2000)
51. Zalesky, A., Vu, H.L., Rosberg, Z., Wong, E.W.M., Zukerman, M.: Modelling and performance evaluation of optical burst switched networks with deflection routing and wavelength reservation. In: Proceedings of IEEE INFOCOM, pp. 1864–1871 (2004)
52. Zalesky, A., Vu, H.L., Rosberg, Z., Wong, E.W.M., Zukerman, M.: Stabilizing deflection routing in optical burst switched networks. IEEE Journal on Selected Areas in Communications **25**(6), 3–19 (2007)
53. Zalesky, A., Vu, H.L., Zukerman, M., Rosberg, Z., Wong, E.W.M.: Evaluation of limited wavelength conversion and deflection routing as methods to reduce blocking probability in optical burst switched networks. In: Proceedings of IEEE ICC, pp. 1543–1547 (2004)

Chapter 5
TCP over OBS Networks

5.1 Introduction

The TCP is one of the important protocols of the IP suite. It provides reliable, in-order delivery of a stream of bytes from a program on one computer to a program on another computer thus making it suitable for applications like file transfer and e-mail. Many studies in the Internet have shown that about 90% of the applications in the Internet use the TCP. With ever-increasing number of Internet users and bandwidth-hungry applications, the technology used in the core network has evolved to high-speed optical networks. The design of an efficient and reliable transport protocol in high-speed networks has become a challenging issue due to problems associated with the fundamental design of the TCP.

One of the main problems faced by the TCP in a high bandwidth-delay product network such as OBS network is the slow convergence due to the additive increase multiplicative decrease (AIMD) congestion avoidance mechanism. For example, if the link capacity is 10 Gbps and the RTT on the path is around 50 ms, a TCP connection with an average packet size of 1.5 KB takes about half an hour to attain the maximum window size even with a packet loss probability of 10^{-8} [16]. Next, in an OBS network, since the packets are assembled into bursts at the ingress node, the TCP segments from a single session might be aggregated into one or many bursts depending on the arrival rate of the traffic and the burst formation time. In such a case, a single burst loss might lead to a loss of several TCP segments at the same time. The number of segments lost in a single burst depends on many factors which is not simple to evaluate. Depending on the number of TCP segments lost, the TCP sender might perceive a single burst loss as severe congestion and reduce the congestion window drastically. Further, any contention resolution techniques used in the OBS network, such as the deflection routing and burst-level retransmission, might lead to out-of-order delivery of packets which is equivalent to the loss of packets. So the fundamental problem with data delivery using the TCP over OBS network is apparently due to the AIMD mechanism used by the TCP sender, and the burst aggregation and the random contention losses in the OBS network. Hence, the study of the TCP has been an important issue of research in OBS networks [32].

Although there are several variants of the TCP proposed for high-speed networks, their performance in OBS networks has not been studied. Any such study or the design of new variants of the TCP for OBS networks requires a deeper understanding of the behavior of TCP over OBS networks. The impact of burst aggregation and the burst losses due to contention on the TCP throughput should be analyzed to design any new methods to improve the performance of the TCP. Evaluating the TCP throughput in OBS networks analytically helps to understand the effect of increased end-to-end delay and the random burst losses on the throughput. This chapter discusses the major factors which affect the performance of the TCP in OBS networks. Then various mathematical models that analyze the behavior of the TCP over OBS networks are presented. These models help not only to evaluate the dependence of the throughput on various parameters of OBS network but also to indicate the approach to design better models for the TCP over high-speed networks in which the losses are correlated. This chapter also discusses some variants of the TCP proposed for OBS networks to improve the throughput by considering the nature of losses in these networks.

5.2 A Brief Introduction to TCP

The TCP is a reliable, connection-oriented protocol which provides *streaming* service to the applications. The TCP sender (henceforth called TCP or sender) forms a packet called a *TCP segment* by gathering some bytes of the data from the application and sends it to the TCP receiver (simply called receiver). The TCP segment is encapsulated in an IP datagram along with the TCP header and payload. TCP specifies a maximum segment size (MSS) which is the largest amount of data, in bytes, that TCP is willing to send in a single segment. Throughout this chapter, the terms packet and segment are used interchangeably to refer to an IP packet with a TCP segment.

The main function of the TCP is congestion control which is done by sender by adjusting the number of segments sent in each round to prevent the buffer overflow at the receiver as well as at the intermediate nodes. This is done by using a window control variable called congestion window (or simply window) at the sender. If the TCP is aware of the available buffer size at the bottleneck, the window could be set to this value and the actual transmission window could be set to the minimum of the window advertised by the receiver (receiver window) and congestion window. To ensure reliable data transfer and to detect congestion in the network, the TCP uses acknowledgments from the receiver for each segment sent.

The ACK mechanism is another important feature of the TCP. When a TCP segment is lost, but the subsequent segment is delivered to the receiver, it buffers the new segment and regenerates the ACK for the segments received so far in sequence. This regenerated ACK, called duplicate ACK, indicates the sender about a lost segment. Some variants of TCP make use of these duplicate ACKs to detect lost segments and to start a recovery procedure. For each round of segments transmitted the sender starts a timer after the expiry of which the segments are considered to be lost if they are not acknowledged. If the sender does not receive any ACKs before the timer expires or the window fails to increase beyond 3, it is perceived as serious congestion.

5.2.1 Phases in the Congestion Control

There are different phases in the congestion control mechanism of the TCP which essentially control the rate of transmission in the network to avoid congestion and to retransmit the lost packets. These phases probe the network for available bandwidth and transmit the data in an incremental fashion to avoid congestion. When some packets are lost due to congestion, recovery phase is used to send the lost packets as well as the new packets. These phases also control the size of the congestion window which is used to determine the exact number of segments to be sent in each round. The various phases of congestion control in the TCP are briefly mentioned below.

5.2.1.1 Slow Start

The TCP probes the network slowly for the available capacity by increasing the number of segments sent in each round gradually to avoid congesting the network. The slow start phase is used for this purpose both at the beginning of the TCP connection and when the transfer is resumed after the time-out.

At the beginning of the slow start phase, the window size is set to 1 and it is increased whenever the previous segments are acknowledged. After the first segment is acknowledged, the sender sends two more segments. When these two segments are acknowledged, it sends four segments and this continues. In this phase the congestion window increases exponentially. To control the exponential increase in window forever, the TCP switches to congestion avoidance phase after the window reaches a pre-set value called slow start threshold. The initial value of the slow start threshold may be high (some variants set it as the size of the receiver window), but it is reduced when the TCP experiences congestion. When the window size becomes equal to the slow start threshold, the sender either uses the slow start algorithm or the congestion avoidance algorithm.

5.2.1.2 Congestion Avoidance

After the window reaches the slow start threshold, the TCP implements the congestion avoidance algorithm during which the window growth is linear. During this phase, the window (say of size W) is increased by $1/W$ each time an ACK is received. So the window is increased by one segment at the end of each RTT. The linear increase in window continues either till the maximum window size is reached or till another packet loss is detected. In case the sender detects the loss of packets (indicated by duplicate ACKs) the slow start threshold is set to half of the window size at the time of loss. If no ACKs are received before the expiry of retransmission timer (known as a time-out event), the window size is set to one segment and the TCP returns to the slow start phase. The sender uses the slow start algorithm to increase the window again to the new slow start threshold. Whenever the retransmission timer expires, the sender doubles the value of the retransmission time-out (RTO). This technique of updating RTO is known as *exponential back-off*.

5.2.1.3 Fast Retransmit and Fast Recovery

The TCP receiver sends a duplicate ACK whenever segment arrives out of order to indicate the actual segment expected. A duplicate ACK can be received by the sender due to a number of reasons. It can be due to the lost segments in which case each segment received after the lost segment triggers a duplicate ACK. It can also be due to retransmission of ACKs by the network or even due to the out-of-order delivery of segments by the network. The duplicate ACKs are used by the TCP sender in the fast retransmit algorithm to infer about the segment losses.

When three duplicate ACKs (called triple duplicate ACK) are received by the sender, it is considered as an indication of a lost segment due to network congestion. The sender immediately retransmits the lost packet without waiting for the retransmission timer to expire. After the retransmission of the lost segment, the sender starts the fast recovery algorithm. This is used to send new data until a non-duplicate ACK arrives. The fast recovery algorithm followed depends on the TCP implementation also called variant of the TCP. Three popular loss-based TCP variants are presented in the next section which basically differ in the way they handle the fast recovery phase.

5.2.2 Popular Variants of TCP

Three popular loss-based variants of the TCP, namely Reno, NewReno, and SACK, are presented in this section. All of them consider loss in the network as an indication of congestion. In all the variants, the slow start phase and the congestion avoidance phase are similar. They differ only in the fast retransmit/recovery phase due to which they improve the throughput by reducing the time taken to retransmit all the lost segments. There are other variants such as TCP Vegas and TCP Westwood that use the estimated delay along the path to identify congestion. Earlier TCP implementations like TCP Tahoe used only the expiry of the retransmission timer to detect the packet losses. The loss of a segment will cause the retransmission timer to expire and the TCP starts with the slow start phase again with a window size of one and the slow start threshold set to half of the window size at the time of the loss. Such loss recovery method not only leads to long idle periods in waiting for the retransmission timer to expire but also reduces the throughput significantly.

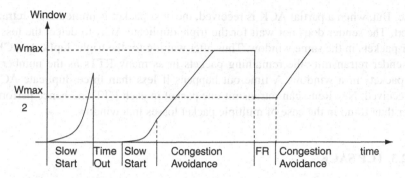

Fig. 5.1 TCP Reno window evolution

5.2.2.1 TCP Reno

Figure 5.1 shows the evolution of the congestion window in the case of TCP Reno. TCP Reno implements the slow start and congestion avoidance algorithm as discussed earlier. It implements the fast retransmit algorithm and avoids the need to wait for a time-out to detect the packet loss unlike the case of Tahoe. On receiving the triple duplicate ACKs, Reno detects the packet loss, retransmits the lost packet, and enters the fast recovery phase. In the fast recovery phase, the slow start threshold is set to half of the window size at the time of loss and the congestion window is set to slow start threshold plus 3. This inflates the congestion window by the number of segments that have reached the receiver (three, due to the triple duplicate ACKs received). For each additional duplicate ACK received, the sender increments the window by 1. During this time the sender can also transmit a new segment if the window size is sufficient. When a new ACK arrives, the sender sets the congestion window to the slow start threshold and exits the fast recovery phase. The TCP sender now continues in the congestion avoidance phase. Such a recovery mechanism gives good performance in the case of occasional packet losses. If multiple packets are lost in a window, the sender invokes the fast retransmit/fast recovery algorithm multiple times but halves the congestion window each time. When multiple packets are lost in a window, the first packet retransmitted causes the receiver to send an ACK to indicate that it expects the second lost packet. Such an ACK for a previously sent but lost packet is called a partial ACK which triggers the fast retransmit/fast recovery phase. Halving the window consecutively for each loss when multiple segments are lost simultaneously leads to a time-out which leads to the poor performance of TCP Reno in this case.

5.2.2.2 TCP NewReno

TCP NewReno modifies the fast retransmit/fast recovery phase of Reno to make it more robust for the case of multiple packet losses in the same window. It implements the fast retransmit algorithm similar to the case of TCP Reno. When a triple duplicate ACK is received, it retransmits the lost packet and enters the fast recovery

phase. But, when a partial ACK is received, the next packet is immediately retransmitted. The sender does not wait for the triple duplicate ACK to detect the loss of
other packets in the same window. Thus after waiting for the triple duplicate ACKs,
the sender retransmits the remaining packets in as many RTTs as the number of
lost packets in a window. A time-out happens if less than three duplicate ACKs
are received. NewReno can retransmit one packet in each RTT and hence performs
better than Reno in the case of multiple packet losses in a window.

5.2.2.3 TCP SACK

Another way to deal with multiple packet losses in a window is to inform the
sender the sequence number of all the packets to be retransmitted by it. This is
done by sending more information with the ACK packets than simply the sequence
number of the next expected packet. Selective acknowledgment (SACK) is a TCP
implementation that uses this feature. If the sender and receiver both implement the
SACK algorithm, and a packet loss is detected with triple duplicate ACKs, the ACK
contains the sequence numbers of the lost packets. The sender retransmits all the lost
packets immediately in response to such an ACK and halves the congestion window
after the retransmission round. All the other phases are similar to that in TCP Reno
and NewReno.

TCP SACK also maintains a variable called pipe which indicates the number of outstanding segments that are in transit. During fast recovery, the sender
sends packets (either new or retransmitted) only when the value of pipe is less
than the congestion window (the number of segments in transit is less than the
window). When the sender sends a packet, the value of pipe is incremented by
one and is decremented by one when it receives a duplicate ACK. It decrements
pipe by two when a partial ACK is received. Similar to NewReno, the fast recovery phase is terminated when an ACK which acknowledges all the packets up to
and including those that were outstanding when the fast recovery phase began is
received.

5.3 Factors Affecting TCP Performance over OBS

The bufferless nature of the OBS networks leads to two major changes in the transmission characteristics in these networks compared to the packet switched networks.
One is the increased end-to-end delay due to burst assembly and signaling and the
other is the random contention losses of bursts even at low traffic loads. Recent
results showed that the delay in the access network and that due to the burst assembly
at the ingress node significantly influence the end-to-end throughput of the TCP [9].
Due to these two features of the OBS networks the behavior of the TCP varies
considerably from that in the Internet [17]. In this section, the impact of the burst
assembly algorithm at the edge node and the random burst losses on the TCP traffic
is studied.

5.3.1 Impact of the Burst Assembly Algorithms

Due to the burst assembly, each TCP segment undergoes an extra delay equal to the burst formation time, which depends on the burst assembly algorithm as well as the input arrival rate. The distribution for the burst formation time for different burst assembly algorithms and input traffic arrival process is presented in Chapter 2. The waiting time of the packets in burst assembly queue as well as the burst transmission queue increases the RTT and the RTO which has a derogatory effect on the throughput of a TCP source. This effect of the additional delay at the edge node on the TCP throughput is called the delay penalty. At the same time, due to the aggregation of several TCP segments in a burst many segments are delivered at once leading to a higher rate of increase in the congestion window. The correlated delivery of segments increases the amount of data sent successfully till a loss occurs due to contention. This effect is called delayed first loss gain or simply correlation benefit which in a way compensates for the delay penalty.

Due to the burst assembly at the edge nodes, there is an increase in the RTT of each TCP segment which can be written as [22]

$$RTT = RTT_0 + 2 \times \tau, \tag{5.1}$$

where τ is the average waiting time of the bursts assumed to be equal at both the edge nodes where a TCP segment as well as its ACK is aggregated into a burst and RTT_0 is the RTT of a burst due to other factors including propagation delay, offset time, processing time, and switching time at each node. Thus, higher the assembly time larger would be the delay penalty. Exact calculation of the delay at the assembly queue is therefore essential to capture the impact of assembly time on the TCP throughput. For time-based assembly algorithm, the delay is fixed by the period of time used for the assembly whereas it depends on the arrival rate of the packets for the size-based assembly algorithm. For a mixed assembly algorithm that uses either time or threshold limit, the analysis is more involved and the burst formation time is given by a probability distribution derived in Chapter 2. Models that evaluate the average waiting time of the bursts at the ingress node that consider the effect of the arrival process as well as the burst assembly algorithm are presented in Chapter 2. If FDLs or any other contention resolution techniques are used at the core nodes, additional delay is added to the above-mentioned RTT. The delay due to the assembly algorithm depends on the threshold values used and the arrival rate of the input traffic. An assembly algorithm that adapts the assembly time according to the current arrival rate was found to provide better throughput at the TCP layer [7, 10, 25, 28].

5.3.2 Impact of Burst Loss

Due to the aggregation of many segments in a burst, a single burst loss leads to a simultaneous loss of many TCP segments. This drastically affects the window growth depending on the number of segments in the burst and the TCP variant. When a burst contains a few TCP segments from a window, its loss triggers the fast retransmission and fast recovery phase without waiting for a time-out in variants such as TCP Reno, NewReno, SACK. In such case the segments lost are retransmitted in one or more rounds depending on the variant and the number of segments lost. However, in low load conditions, a burst might contain all the segments from a window the loss of which leads to the expiry of the retransmission timer. In this case the sender considers the loss as that due to serious congestion and starts from the slow start phase. This kind of identifying a single or few burst losses due to contention with congestion is called false congestion detection. The TCP source senses congestion in the network due to the expiration of the retransmission timer, even though the network is not congested and is not heavily loaded. The time-out in such cases called false time-out (FTO) triggers the slow start phase for each burst loss reducing the throughput of the TCP sender drastically [46]. On the other hand the burst losses may be persistent due to unavailability of bandwidth along the path. In such case, the TCP sender should reduce the congestion window according to the congestion avoidance mechanism. TCP perceives the loss of segments even if they are delivered out of order [1, 26]. Due to the burst assembly algorithms and the contention resolution techniques such as burst-level retransmission, segmentation, and deflection routing, the bursts might deliver TCP segments in different order than that originated at the sender. The frequency of out-of-order delivery depends on the thresholds used for assembly and the QoS criteria used in the aggregation. Performance evaluation of TCP over OBS networks when load balancing techniques are used shows that the throughput is severely affected with burst-level reordering. To avoid this a routing scheme to reduce the differential delay between two paths is proposed which was found to benefit the TCP flows. Bursts are deliberately delayed when sent on shorter paths so that TCP segments reach the destination at the same time on both the paths [1].

TCP variants designed for OBS networks mainly use techniques to inform the sender if the losses detected are due to a single burst loss or due to multiple burst losses. In other words, the sender is indicated if a time-out is an FTO or it is really due to congestion. The TCP sender is given additional information on the state of the network to identify the reason for the loss. For example, machine learning techniques at the edge node can be used to categorize the losses into contention losses and losses due to congestion. The TCP throughput is improved by avoiding the reduction of congestion window for contention losses [20]. These techniques to improve the performance of the TCP also called as TCP variants for OBS networks are discussed in Section 5.5. Such variants of TCP require additional signaling efforts between the sender and the edge or core nodes of the OBS network. The throughput of TCP in OBS networks is also found to be improved due to the use of other techniques including the use of contention resolution techniques at the core nodes. Such techniques are presented in Section 5.6. Modeling the behavior of the TCP over OBS networks not only helps to understand the performance of TCP but also helps to design techniques to improve the performance of TCP. The next section presents different mathematical models for TCP over OBS networks which help to understand the impact of the burst losses and the burst assembly on the throughput.

5.4 Analytical Models for TCP over OBS

Exact characterization of the impact of burst losses on the throughput is difficult due to the lack of good models for the burst losses in OBS networks and also due to the dependence of the BLP on many factors. However, the models presented in this section are fairly accurate to analyze the TCP with some assumptions on the window evolution and the number of segments lost due to a burst loss. The main challenge in characterizing the impact of a burst loss on the throughput is to evaluate the number of segments from a TCP flow in that burst. The average delay suffered by the TCP segments in an OBS network also requires accurate modeling of the edge node to understand its effect on the RTT and RTO values. Due to the interplay of the delay penalty and the correlation benefit, it is difficult to analyze the throughput of a single TCP flow when the traffic has multiple TCP flows, load due to other background traffic, and by considering the other components in the OBS network [13].

Analytical models to evaluate the TCP throughput over OBS networks can be classified into two main categories: those that use Markov regenerative process theory or the renewal theory and those that use the theory of Markov chains. The renewal theory-based models are simple to arrive at and also provide elegant expressions for the throughput in terms of the RTT and the BLP. On the other hand the Markov chain-based models evaluate the impact of many other factors accurately but can be solved only using numerical techniques. This section discusses both these types of models for the TCP in OBS networks. The popular loss-based TCP variants which are modeled include TCP Reno, NewReno, and SACK. The performance of each of these variants differs due to the different loss recovery method adopted by each variant. This section also presents the analytical models for the TCP variant designed for high bandwidth-delay product networks, high-speed TCP, and also the delay-based variant TCP Vegas.

5.4.1 Markov Regenerative Process Theory-Based Models

A major breakthrough in modeling the TCP throughput in the Internet occurred with the popular renewal theory-based model for TCP Reno proposed by Padhye et. al. [24]. This model captures the effect of the fast retransmit and time-out phases on the TCP throughput. Despite many approximations, it gives an elegant closed form expression for the throughput in terms of the packet loss probability and the RTT. There are many analytical models for the TCP throughput in OBS networks based on Padhye's model which study the impact of the burst assembly algorithms and the random losses in the bufferless OBS network on the TCP throughput [11, 44, 45,

47]. The authors in [45] study the performance of TCP over OBS networks by first considering only one TCP flow and then multiple flows with an exponential burst loss model for OBS networks proposed in [44]. The evolution of the congestion window is modeled in terms of rounds in which a round denotes the back-to-back transmission of a window of packets. The round ends when the first ACK is received for the window of packets or when the retransmission timer expires. The duration of a round is assumed to be one RTT and it is assumed to be independent of the window size. Also, the number of packets lost in one round is considered to be independent of that in the other rounds.

The steady-state throughput of a TCP flow can be written in terms of the loss ratio p and the RTT in an OBS network as

$$\text{Throughput} = \frac{1}{\text{RTT}_0 + 2T_b} \sqrt{\frac{S}{2bp}} + o(\frac{1}{\sqrt{p}}), \tag{5.2}$$

where S is the average burst size in terms of the number of packets, RTT_0 is the round trip time in the network, T_b is the burst formation time, and b is the number of ACKed rounds before the window size is increased.

Consider the setup for the TCP connection used in this model as shown in Fig. 5.2. The access network is assumed to be lossless with a bandwidth of B_a (called access bandwidth) and the delay suffered in the access network is assumed to be d_a. The path in the OBS network is assumed to have a propagation delay T_p and the losses along the path follow a Bernoulli distribution with parameter p_b. The transmission delay of the TCP segments and delay due to deburstification are neglected in this model. The reverse path is considered to be lossless and there is no burst assembly or disassembly in the reverse path.

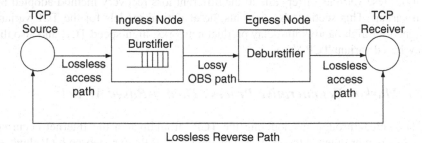

Fig. 5.2 Connection model

Let W_m be the maximum window size advertised by the receiver, L_p be the size of the burst, and T_b be the burst formation time. The TCP sources in this model are classified according to the access bandwidth into fast, slow, and medium classes such that the following inequalities are satisfied:

$$\text{Slow source}: \quad \frac{B_a T_b}{L_p} \le 1, \tag{5.3}$$

$$\text{Fast source}: \quad \frac{B_a}{W_m L_p} T_b \ge 1, \tag{5.4}$$

$$\text{Medium source}: \quad 1 < \frac{B_a T_b}{L_p} < W_m. \tag{5.5}$$

A fast class TCP source is characterized by an access bandwidth (B_a) high enough to let all the segments from a window be aggregated into a single burst. On the other hand, a slow class TCP source is such that a burst contains at most one segment from the TCP flow. The medium class sources fall in between these two classes. Such a classification of TCP source enables the number of segments in a burst to be evaluated accurately but the model developed with these assumptions is incapable of studying the performance of a generic speed TCP source. This drawback has been fixed in the renewal theory-based model for TCP presented in the subsequent sections.

The throughput of the TCP connection in a time interval $[0, t]$ is defined as $B_t = \frac{N_t}{t}$, where N_t is the number of packets sent in this time. The steady-state long-term throughput of the TCP source is then defined as

$$B = \lim_{t \to \infty} B_t = \lim_{t \to \infty} \frac{N_t}{t}. \tag{5.6}$$

The evolution of the TCP window is defined as a Markov regenerative (renewal) process $\{W_i\}_i$, where W denotes the window size. A regenerative process is a stochastic process with the property that there exist time points at which the process (probabilistically) restarts itself. Let S_i denote the ith renewal period in the evolution of the congestion window. Let M_i be the reward received in each regenerative period which is the number of packets sent in a renewal period. Then using the renewal theorem the average reward per unit time is equal to the throughput of the TCP source which can be defined as

$$B = \frac{E[M]}{E[S]}. \tag{5.7}$$

Using the renewal theorem, the throughput of a TCP flow can be computed by taking the ratio of the number of packets sent during a cycle of window growth to the average duration of the cycle. This procedure is used to compute the throughput of a slow and a fast class TCP source.

5.4.1.1 Slow Class TCP Sources

With a slow class TCP source, each burst contains at most one segment. So the throughput can be modeled similar to that in the case of packet switched networks. Figure 5.3 shows the evolution of the congestion window in a renewal cycle

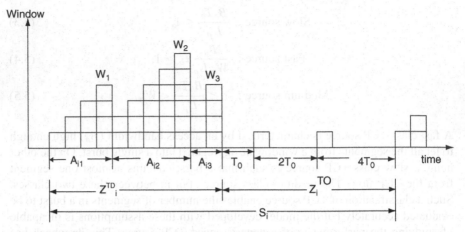

Fig. 5.3 Evolution of congestion window for slow class TCP sources

denoted by S_i consisting of both triple duplicate (TD) and time-out (TO) losses. This evolution ignores the slow start and fast recovery phases assuming that most of the packets are sent only during the congestion avoidance phase. Therefore, the window size W is increased by $1/W$ on the reception of each ACK (increases by 1 after each successful sending round). The case of delayed ACKs is neglected here so that an ACK is sent for every segment received. Each burst loss leads to the loss of only one segment. As the burst loss events are independent and Bernoulli distributed, the segment losses are also statistically independent.

Let Z_i^{TO} and Z_i^{TD} denote the duration of a sequence of time-outs and triple duplicate periods (TDPs). Let n_i be the number of TDPs in the interval Z_i^{TD} which is 3 as shown in the figure. Let Y_{ij} be the number of segments sent during the jth TDP of the interval Z_i^{TD} of duration A_{ij} and let R_i be the number of segments sent during the Z_i^{TO} period. Let W_i be the window size at the end of the ith TDP. With these definitions, the total number of segments sent during the ith renewal period denoted by M_i and the duration of each renewal period S_i can be written as

$$M_i = \sum_{j=1}^{n_i} Y_{ij} + R_i \qquad (5.8)$$
$$= Y_i + R_i,$$

$$S_i = Z_i^{TD} + Z_i^{TO} \qquad (5.9)$$
$$= \sum_{j=1}^{n_i} A_{ij} + Z_i^{TO}.$$

Since the occurrence of TDP and TO periods are independent of each other, the average number of packets sent during the period S_i and the average duration of the period can be written as

$$E[M] = E[\sum_{j=1}^{n_i} Y_{ij}] + E[R], \tag{5.10}$$

$$E[S] = E[\sum_{j=1}^{n_i} A_{ij}] + E[Z_i^{\text{TO}}].$$

On assuming $\{n_i\}_i$ to be a sequence of iid random variables independent of A_{ij} and Y_{ij},

$$E[\sum_{j=1}^{n_i} Y_{ij}] = E[n]E[Y], \tag{5.11}$$

$$E[\sum_{j=1}^{n_i} A_{ij}] = E[n]E[A].$$

Hence substituting the average values of M and S given above in Eq. 5.7

$$B = \frac{E[Y] + Q * E[R]}{E[A] + Q * E[Z^{\text{TO}}]}, \tag{5.12}$$

where Q is the probability of a TDP ending with a TO. Note that $Q = \frac{1}{E[n]}$ as there is only one TO out of n_i loss indications during Z_i^{TD}.

To evaluate the throughput B of a TCP source in this setting, the exact number of segments sent in each of the TDP and TO periods along with their average duration must be evaluated. Consider only the TDP in a renewal cycle. Figure 5.4 shows the window evolution during an ith TDP (A_i as shown in Fig. 5.3) which starts immediately after a TD loss indication.

At the start of the ith TDP, the window is size is given by $W_{i-1}/2$ where W_{i-1} is the window size at the end of $i-1$th TDP. Let X_i be the round when the first packet loss occurs and α_i denote the count of the first packet lost. After the packet α_i, $W_i - 1$ more segments are sent before the end of the TDP. Thus the total number of packets sent in the ith TDP, $Y_i = \alpha_i + W_i - 1$. Hence,

$$E[Y_i] = E[\alpha] + E[W] - 1. \tag{5.13}$$

The expected number of packets sent in a TDP up to the first loss can be written as

$$E[\alpha] = \frac{1}{p_b}, \tag{5.14}$$

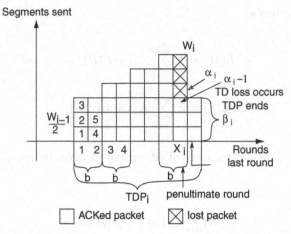

Fig. 5.4 Packets sent during TDP$_i$

where p_b is the packet loss probability or burst loss probability (each burst consists of only a single packet for slow sources). Also the duration of the ith TDP, A_i, is given by

$$E[A] = E[\text{RTT}](E[X] + 1) \tag{5.15}$$

which follows from the fact that a TDP period consists of $X_i + 1$ rounds each of duration RTT$_{ij}$. RTT$_{ij}$ denotes the duration of the jth round of ith TDP. To derive $E[X]$ and $E[W]$, consider the evolution of congestion window in terms of a number of rounds. Since window increases linearly in the congestion avoidance phase, the window size at the end of a TDP denoted by W_i can be expressed as

$$W_i = \frac{W_{i-1}}{2} + X_i - 1. \tag{5.16}$$

Then the number of segments sent during the ith TDP, Y_i can be written as

$$Y_i = \sum_{k=0}^{X_i-1} \left(\frac{W_{i-1}}{2} + k\right) + \beta_i, \tag{5.17}$$

where β_i is the number of segments sent in the last round assumed to be uniformly distributed between 1 and $W_i - 1$. Thus $E[\beta] = E[W]/2$. Also, assuming X_i and W_i to be mutually independent and using Eqs. 5.13, 5.14, 5.16, and 5.17, the mean window size at the end of a TDP, $E[W]$ can be written as

$$E[W] = \sqrt{\frac{8}{3p_b}} + o(1/\sqrt{p_b}). \tag{5.18}$$

Using the expression for $E[W]$, $E[X]$ can be evaluated from Eq. 5.16. $E[X]$ is in turn substituted in Eq. 5.15 to calculate $E[A]$.

If losses are only TD losses and there are no time-outs, the expression for TCP throughput becomes

$$B = \frac{1}{\text{RTT}}\sqrt{\frac{3}{2p_b}} + o(1/\sqrt{p_b}). \tag{5.19}$$

To consider time-outs also in the renewal cycle shown in Fig. 5.3, the expected number of packets sent in the time-out states, i.e., $E[R]$ and the expected time spent in the TO states, i.e., $E[Z^{\text{TO}}]$ have to be determined. To calculate $E[R]$ the distribution of number of time-outs in a time-out sequence is required. A sequence of time-out states is always followed by a successful packet reception at the receiver. Hence, the number of time-outs in a time-out sequence follows a geometric distribution with parameter p_b so that

$$\text{Prob}[R = k] = p_b^{k-1}(1 - p_b) \tag{5.20}$$

$$E[R] = \frac{1}{1 - p_b}. \tag{5.21}$$

Further, using the principle of exponential back-offs during consecutive time-outs, the duration of a sequence of k time-outs is given by

$$H_k = \begin{cases} (2^k - 1)\text{RTO}, & k \leq 6 \\ (63 + 64(k - 6))\text{RTO}, & k \geq 7. \end{cases} \tag{5.22}$$

Therefore, the average duration of the TO period is given by

$$E[Z^{\text{TO}}] = \sum_{k=1}^{\infty} H_k \text{Prob}[R = k] \tag{5.23}$$

$$= \text{RTO}\frac{1 + p_b + 2p_b^2 + 4p_b^3 + 8p_b^4 + 16p_b^5 + 32p_b^6}{1 - p_b}.$$

The throughput for a slow class TCP source can be obtained by substituting for all the terms in Eq. 5.12 which includes losses due to time-outs and triple duplicate ACKs and also a limit for the maximum window. The throughput of a slow class TCP source is given by

$$B^s(W_m, \text{RTT}, p_b, \text{RTO}) = \begin{cases} \dfrac{\frac{1-p_b}{p_b}+E[W]+\hat{Q}(E[W])\frac{1}{1-p_b}}{\text{RTT}(\frac{E[W_u]}{2}+1)+\hat{Q}(E[W])\text{RTO}\frac{\hat{f}(p_b)}{1-p_b}}, & \text{for } E[W_u] < W_m \\[4mm] \dfrac{\frac{1-p_b}{p_b}+W_m+\hat{Q}(W_m)\frac{1}{1-p_b}}{\text{RTT}(\frac{W_m}{8}+\frac{1-p_b}{p_b W_m}+2)+\hat{Q}(W_m)\text{RTO}\frac{\hat{f}(p_b)}{1-p_b}}, & \text{otherwise}, \end{cases}$$

$$\tag{5.24}$$

where

$$E[W_u] = 1 + \sqrt{\frac{8(1 + p_b)}{3p_b} + 1},$$ (5.25)

$$\hat{Q}(u) = \min(1, \frac{3}{u}),$$

$$f(p_b) = 1 + p_b + 2p_b^2 + 4p_b^3 + 8p_b^4 + 16p_b^5 + 32p_b^6.$$

The authors in [12] also model TCP Reno flows in a similar way for the slow class TCP sources but use Markov chain-based model for the case of fast sources by considering the slow start phase also.

5.4.1.2 Fast Class TCP Sources

The growth of the congestion window is modeled as a sequence of rounds similar to the case of slow sources. According to the definition of a fast source in Eq. 5.4, the entire window of segments in a round are aggregated into a single burst. When a burst is lost the entire window of segments are lost and the sender goes to the time-out state. Such a round is called lossy round. The round in which the burst successfully reaches the receiver is called a successful round. Figure 5.5 shows a renewal cycle during the evolution of congestion window for fast class TCP sources. It can be simply written as a collection of sequence of successful rounds when the bursts are delivered successfully followed by a sequence of lossy rounds when the bursts are lost. The ith renewal cycle is called the ith time-out period (TOP) (shown as TOP_i in the figure). Due to the losses leading to time-out state, the fast retransmit and fast recovery phases are not activated for the case of fast TCP sources. On every burst loss, the sender resets the window size to 1 and starts from the slow start phase in which it tries to retransmit all the segments lost in the previous round. For every consecutive lossy round, the sender doubles the RTO till it reaches 64*RTO.

During the period TOP_i, let Y_i be the number of segments sent in a sequence of the successful rounds with a total duration of A_i, H_i be the number of segments sent after the first time-out loss, and Z_i^{TO} be the duration of the sequence of lossy rounds. Using the definition of the throughput in Eq. 5.7, for a fast TCP source

$$B^f = \frac{E[Y] + E[H]}{E[A] + E[Z^{TO}]}.$$ (5.26)

The average duration of the TO period, $E[Z^{TO}]$, is same as that given in Eq. 5.23. $E[R]$ evaluated in Eq. 5.21 gives the expected number of time-out losses in a renewal cycle. The average number of segments sent after the first time-out is thus given by

$$E[H] = E[R] - 1 = \frac{p_b}{1 - p_b}.$$ (5.27)

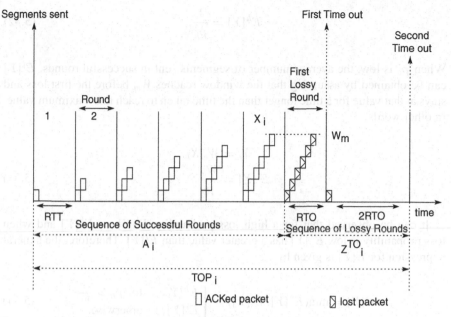

Fig. 5.5 Evolution of congestion window for fast class TCP sources

Let X_i denote the number of successful rounds in the ith TOP so that

$$E[A] = E[X]\text{RTT}. \tag{5.28}$$

Since a Bernoulli model is assumed for the burst losses, the random variable X is geometrically distributed with the parameter p_b. Therefore,

$$E[X] = \frac{1 - p_b}{p_b}. \tag{5.29}$$

To compute $E[Y]$, two scenarios are considered: one when the BLP is high and the other when it is low. When p_b is high, the limit on the maximum window size can be neglected (because the window never reaches the maximum value) and also the slow start phase can be neglected. In such a case, the window always increases linearly which can be expressed as

$$W_i = X_i + 1, \tag{5.30}$$

$$Y_i = W_i \left(\frac{W_i + 1}{2} \right). \tag{5.31}$$

Therefore $E[Y]$ when p_b is high denoted by $E^h[Y]$ can be evaluated using Eqs. 5.29, 5.30, and 5.31 as

$$E^h[Y] = \frac{1}{p_b^2}. \tag{5.32}$$

When p_b is low, the average number of segments sent in successful rounds, $E^l[Y]$ can be obtained by assuming that the window reaches W_m before the first loss and stays at that value for a time longer than the time taken to reach this maximum value. In other words,

$$Y_i = W_m X_i,$$
$$E^l[Y] = \frac{W_m}{p_b}. \tag{5.33}$$

It can be observed that for a high loss probability $E^l[Y] > E^h[Y]$ and when loss probability is low $E^h[Y]$ has a greater value than $E^l[Y]$. Therefore, the general expression for $E[Y]$ is given by

$$E[Y] = \min(E^h[Y], E^l[Y]) = \begin{cases} E^h[Y], & \text{for } p_b > \frac{1}{W_m} \\ E^l[Y], & \text{otherwise.} \end{cases} \tag{5.34}$$

The throughput of a fast TCP source is then given by

$$B^f = \begin{cases} \frac{p_b^3 - p_b + 1}{RTT[p_b(1-p_b)^2 + p_b^3 f(p_b)]}, & \text{for } p_b > \frac{1}{W_m} \\ \frac{W_m - p_b W_m + p_b^2}{RTT[(1-p_b)^2 + p_b^2 f(p)]}, & \text{otherwise.} \end{cases} \tag{5.35}$$

Though the throughput of a medium source cannot be evaluated in a similar manner, it is obvious that the throughput of such a source lies in between that of B^s and B^f, i.e.,

$$B^s < B^m < B^f. \tag{5.36}$$

In such a model the throughputs B^s and B^f are independent of the access bandwidth whereas B^m depends on the access bandwidth. By comparing the TCP throughput in the presence and absence of burstifiers, *correlation benefit* and *delay penalties* can be quantified. Delay penalty degrades the throughput of a TCP source by a factor P_d which is proportional to the ratio of the burst formation time to the round trip time in the absence of the burstifier. Hence,

$$P_d = \frac{RTT_0 + T_b}{RTT_0} = \frac{RTT}{RTT_0}. \tag{5.37}$$

The correlation benefit for slow and fast sources is defined as

$$C_b^s = \frac{B_0^s}{B_{\text{NB}}}, \tag{5.38}$$

$$C_b^f = \frac{B_0^f}{B_{\text{NB}}}, \tag{5.39}$$

where C_b^s and C_b^f denote the correlation benefit for slow and fast sources, respectively, and B_{NB} is the TCP throughput when the burstifier and deburstifier are missing. B_0^s is the TCP throughput in the absence of burstifier and B_0^f denotes the TCP throughput when $\frac{T_b}{\text{RTT}_0} = 0$. But B_0^f cannot be considered as the throughput in the absence of burstifier as the assumptions used for fast source (a single burst contains all the segments in a window) do not hold if the burstification process is neglected. If C_b^m denotes the correlation benefit for a medium source, it is easy to observe from Eq. 5.36 that

$$C^s < C^m < C^f. \tag{5.40}$$

From Eq. 5.38, it can be seen that $C_b^s = 1$. Therefore a TCP source belonging to slow class does not have any correlation benefit. Correlation benefit for the fast and medium sources increases with more number of segments aggregated in a burst. It can also be seen that the correlation benefit is maximized for $p_b = 1/W_m$ and it is negligible for very low or very high values of p_b.

5.4.2 Renewal Theory-Based Model for Loss-Based Variants

The model discussed in the previous section provides a closed form expression for the TCP throughput only for the cases of slow and fast sources. As mentioned earlier, such a classification of the sources based on the access bandwidth does not model the throughput of TCP independent of the bandwidth. In this section, the renewal theory-based model is extended for any generic source for the popular loss-based TCP variants TCP SACK, NewReno, and Reno by considering the effect of burst assembly and burst losses on the throughput [8, 27, 47].

In this section, the throughput of only a single TCP flow is considered which works on the assumption that the BLP, p_b, is insensitive to the input arrival rate into the network. Let λ_a be the rate of arrival of TCP segments at the ingress node which depends on the access bandwidth and S be the number of TCP segments in a burst (average value to keep it independent of the burst assembly algorithm). The variables Y_i, X_i, A_i, H_i, and Z_i^{TO} mean the same as in the previous section. The burst size S is at least 1 even when the burst formation time $T_b = 0$ and is at the most equal to the maximum window size W_m. Under these assumptions, the burst size is given by

$$S = \min(\lambda_a T_b + 1, W_m). \tag{5.41}$$

In this section, the delayed ACK mechanism is assumed to be present so that the window is increased for every b acknowledged rounds.

5.4.2.1 TCP SACK

TCP SACK uses a field in the ACK packets which indicates the lost packets in each burst which helps faster recovery of the lost packets. There are two types of loss indication techniques used in SACK: one is by the triple duplicate ACKs and the other one is by time-outs. Hence, the throughput of a SACK TCP source can be expressed using Eq. 5.12.

Figure 5.6 shows the retransmission process of packets that are lost in a TDP. All the packets lost in a round are due to the loss of a single burst and these are retransmitted all at once in the next round along with some new packets. The total number of packets in a round is determined by the current window size as shown in the figure.

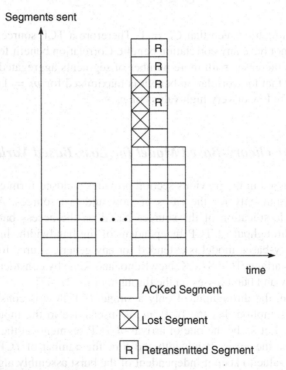

Fig. 5.6 TCP SACK retransmission

Let $(\beta_i + 1)$th burst be the first burst lost in TDP_i in which the first segment is $(\alpha_i + 1)$ and let W_{X_i} be the window size at the end of TDP_i. Let γ_i be the number of additional segments sent in the same round after the burst loss. When three duplicate ACKs are received, the TCP source immediately retransmits the lost segments (say

S in number when S is the burst size) along with $W_{X_i} - S$ new segments. Therefore the number of segments sent in a TDP, Y_i, can be written as $Y_i = \alpha_i + \gamma_i + W_{X_i} - S$ and by approximating $E[\gamma_i]$ with $E[W_X]/2$ gives

$$E[Y] = E[\alpha] + \frac{3}{2}E[W_X] - S. \tag{5.42}$$

Since the burst losses are assumed to be independent of each other, the random variable β_i follows a geometrical distribution. Further, $\alpha_i = S\beta_i$ so that $E[\alpha]$ is given by

$$E[\alpha] = SE[\beta] = \frac{S}{p_b}. \tag{5.43}$$

Therefore $E[Y]$ is given by

$$E[Y] = \frac{3}{2}E[W_X] + \frac{1 - p_b}{p_b}S. \tag{5.44}$$

To evaluate the average window size at the end of the TDP, two cases are considered. The first case is when the W_m is so large that most of the time the window size $W_X < W_m$. The second case is when $W_X = W_m$ during most of the rounds. This happens when the burst loss probability is very low. When $W_X < W_m$, examining the evolution of window during the congestion avoidance phase similar to that explained when Eq. 5.16 is obtained,

$$W_{X_i} = \frac{W_{X_{i-1}}}{2} + \frac{X_i}{b} \tag{5.45}$$

from which the average value can be written as

$$E[X] = \frac{b}{2}E[W_X]. \tag{5.46}$$

The number of segments sent in TDP$_i$ can be computed by adding the number of segments sent in different rounds as

$$Y_i = \sum_{k=0}^{X_i/b-1} (\frac{W_{X_{i-1}}}{2} + k)b + W_{X_i} - S, \tag{5.47}$$

where the number of rounds is now only X_i/b because every b rounds are acknowledged once. Therefore,

$$E[Y] = \frac{3bE[W_X]^2}{8} + (1 - \frac{b}{2})E[W_X] - S. \tag{5.48}$$

From the above equations, $E[W_X]$ can be written as

$$E[W_X] = \frac{2(b+1)}{3b} + \sqrt{\frac{2b+2^2}{3b} + \frac{8S}{3bp_b}}. \qquad (5.49)$$

Then $E[Y]$ can be evaluated using Eqs. 5.49 and 5.48. Also by substituting for $E[W_X]$ in Eq. 5.46, $E[X]$ can be evaluated from which the expected duration of TDP_i, $E[A]$, can be evaluated using Eq. 5.15.

When $W_X = W_m$, Eq. 5.45 cannot be used to estimate the number of rounds in a TDP. As window has a size of W_m most of the time, $E[Y]$ is evaluated by setting $E[W_X] = W_m$ in Eq. 5.44 to arrive at

$$E[Y] = \frac{3}{2}W_m + \frac{1 - p_b}{p_b}S. \qquad (5.50)$$

The number of packets successfully sent before a TD loss event is S/p and the window grows from $W_m/2$ to W_m in each TDP. By equating the total number of segments sent during the congestion avoidance phase to S/p, $E[X]$ can be evaluated. From $E[X]$, $E[A]$ is evaluated using Eq. 5.15.

Since $W_X = W_m$, there could also be TO losses which can be modeled similar to that explained in the previous section. $E[H]$ is evaluated using Eq. 5.27 and $E[Z^{TO}]$ using Eq. 5.23. It is assumed that a time-out loss always occurs when all the segments in the last round (X_i of them) of a TDP are lost. Probability of such an event happening is given by

$$P(W_X) = p_b^{\frac{W_X}{S} - 1}. \qquad (5.51)$$

Therefore a TDP will always be followed by a time-out with probability $P(W_X)$ or by another TDP with probability $1 - P(W_X)$. The expected ratio of the probability of a time-out to the probability of a triple duplicate loss is given by

$$Q(E[W_X]) = E[P(W_X)] \simeq p_b^{\frac{W_X}{S} - 1}. \qquad (5.52)$$

Finally, substituting the expressions for $E[Y]$, $E[A]$, $E[H]$, $Q = Q(E[W_X])$, and $E[Z^{TO}]$ in Eq. 5.12 gives the throughput of a SACK source to be

$$B = \frac{1}{RTT_0 + 2T_b}\sqrt{\frac{3S}{2bp_b}} + o\left(\frac{1}{\sqrt{p_b}}\right). \qquad (5.53)$$

The actual RTT of a TCP flow is increased by $2T_b$ because of the burst assembly in both the directions when the burst formation time at the edge node is T_b.

5.4.2.2 TCP NewReno

Unlike SACK, NewReno retransmits at most one segment lost from a burst after receiving each partial ACK. The number of rounds in the fast retransmit phase is equal to the number of packets lost (which is the size of the burst S). The total number of packets sent in a TDP is same as that in the case of SACK. Therefore, the throughput of TCP NewReno for a low value of p_b and when $W_X < W_m$ is given by

$$B = \frac{S/p_b}{\text{RTT}\left(\sqrt{\frac{2bS}{3p_b}} + S\right)} + o\left(\frac{1}{\sqrt{p_b}}\right). \tag{5.54}$$

5.4.2.3 TCP Reno

In TCP Reno, the fast retransmit/fast recovery phase is different from that in SACK and NewReno. Consider the case when there are only losses indicated by triple duplicate ACKs. The burst length S and the loss probability p_b are assumed to be small so that Reno can recover from multiple packet losses by retransmitting them in multiple rounds. Assume that the congestion window is always greater than 3 so that triple duplicate ACKs can be received for each packet retransmitted, i.e., $\log W_X > S$. Figure 5.7 shows the retransmission of packets by Reno with only triple duplicate losses. The sender halves the window after the retransmission of each lost segment. There are S additional rounds required to retransmit the S packets compared to the case of NewReno.

The number of new segments sent along with the retransmitted packets during the retransmission rounds can be written as

$$Y_a = \sum_{k=0}^{S-1} \frac{1}{2^k} W_{X_i} - S, \tag{5.55}$$

where X_i is the number of packets sent before the first TD loss. Thus the total number of segments sent during TDP_i is

$$Y_i = \alpha_i + \gamma_i + Y_a. \tag{5.56}$$

Once a new ACK arrives, TDP_{i+1} starts with a sending window size of $W_{X_i}/2^S$. The sending window size is therefore given by

$$W_{X_i} = \max\left(1, \frac{W_{X_{i-1}}}{2^S}\right) + \frac{X_i}{b}. \tag{5.57}$$

Since the congestion window size must remain greater than 3 in order to trigger fast recovery in Reno,

$$W_{X_i} = \frac{W_{X_{i-1}}}{2^S} + \frac{X_i}{b}. \tag{5.58}$$

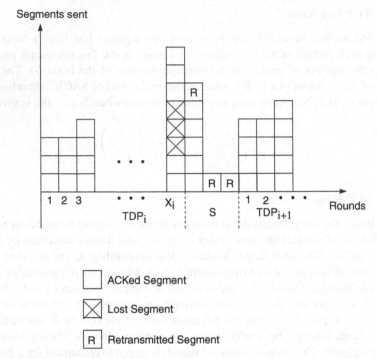

Fig. 5.7 TCP Reno window evolution with only TD loss indication

The remaining parameters needed to compute the throughput can be obtained similar to the case of SACK.

Next, consider a renewal period in the evolution of the congestion window with Reno which consists of both TDP and TOP. Figure 5.8 shows two TDPs separated by a TO period along with the recovery phase.

When burst length S is large TCP Reno may end up in a time-out state due to lack of required number of duplicate ACKs to trigger the fast retransmit phase. This happens in the case of Reno because it halves the congestion window after the retransmission of each lost segment and the window size eventually falls below 3 so that the sender waits till the retransmission timer expires. In such a case losses could be indicated by both triple duplicate ACKs and time-outs.

If the window size becomes smaller than 3 before all the packets are retransmitted, $\log W_X < S$. In this case a TDP is followed by $\log W_X - 1$ retransmissions and then a TOP. The next TDP starts with a window size of 1, i.e., the slow start phase. In this case the average number of segments transmitted is given by

$$E[Y] = \left(\frac{5}{2} - \frac{1}{2^{\log W_x}}\right) E[W_X] - \log W_X + \frac{S}{p_b} \qquad (5.59)$$

and

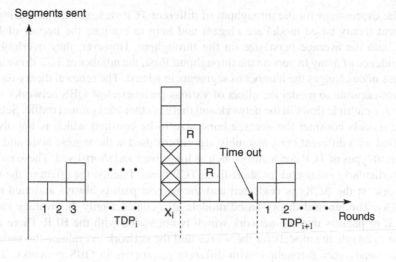

Fig. 5.8 TCP Reno window evolution with TO and TDP

$$E[X] = bE[W_X]. \qquad (5.60)$$

Using the relationship between $E[Y]$, $E[X]$, and $E[W_X]$, the other parameters required to evaluate the throughput can be evaluated. The steady-state throughput for TCP Reno for the case of $\log W_X < S$ is given by

$$B = \frac{\frac{S}{p_b}}{\text{RTT}\left(\sqrt{\frac{2bS}{p_b}} + \log\sqrt{\frac{2S}{bp_b}} + \text{RTO}\right)} + o\left(\frac{1}{\sqrt{p_b}}\right). \qquad (5.61)$$

and for the case of $\log W_x > S$ it can be obtained on similar lines as

$$B = \frac{\frac{S}{p_b}}{\text{RTT}\sqrt{\frac{2bS}{p_b}} + S} + o\left(\frac{1}{\sqrt{p_b}}\right). \qquad (5.62)$$

The expressions for the throughput of different TCP variants obtained using the renewal theory-based model are elegant and help to evaluate the impact of BLP, RTT, and the average burst size on the throughput. However, they overlook the dependence of many factors on the throughput. First, the number of TCP flows at the ingress node changes the number of segments in a burst. The renewal theory models are not accurate to model the effect of various parameters of OBS networks when there are multiple flows in the network and there is other background traffic. Second, these models consider the average burst size to be constant which is not always justified with different burst assembly algorithms used at the ingress node and with different types of TCP connections such as long-lived and short-lived. These models are particularly not useful for short-lived TCP flows. Finally, the effect of the OBS network on the ACKs is neglected and the reverse path is always assumed to be lossless. The renewal theory-based models also consider only the average rate of arrival of packets into the network which is not varied with the BLP. There is no scope to couple the models for the source and the network to evaluate the variation in the steady-state throughput with different parameters in OBS networks. These observations motivate the proposal of Markov chain-based models for TCP over OBS network presented in the next section.

5.4.3 Markov Chain Theory-Based Models

As mentioned in the previous section, the renewal theory-based models are useful mainly with only time-based assembly algorithms at the ingress node. This is due to the classification of sources based on the access bandwidth which enables the definition of average burst size. The number of packets lost in a burst are considered at only the extreme ends, i.e., either the burst contains all the segments in a window or it contains only one segment. The packet loss probability which is used in the expression for throughput of TCP is assumed to be same as the BLP. All these assumptions are invalid when there are multiple flows with correlation among them and when a mixed assembly algorithm (for example, the MBMAP algorithm described in Chapter 1) is used. When the MBMAP algorithm is used for assembly, the burst size is variable and thus the number of packets belonging to a TCP flow in a burst is also a random variable. Further, the models in the previous section assume the BLP to be independent of the burst size distribution or other parameters of OBS network. Using the renewal theory-based models for TCP over OBS network makes it difficult to consider the impact of various components of the network on the throughput explicitly. In this section a model for a generic TCP source as well as the network components based on Markov chain theory is presented which couples the TCP model with the network model [2, 3].

Figure 5.9 gives a high-level view of the model. Traffic from multiple TCP sources arrives at the ingress node of an OBS network to get aggregated into bursts. There are two subsystems in the model. The source model gives the average rate of arrival into the network which also gives the cumulative throughput due to all the

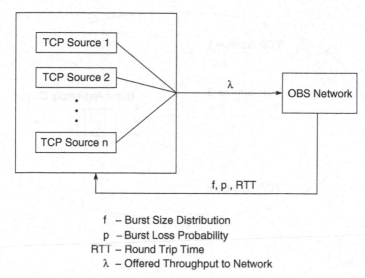

f – Burst Size Distribution
p – Burst Loss Probability
RTT – Round Trip Time
λ – Offered Throughput to Network

Fig. 5.9 High-level view of the model

sources. The network model computes the BLP, end-to-end delay in the network, and burst size distribution which is again used by the source model to compute the throughput. Evaluation of the throughput at the source, BLP, and end-to-end delay in the network is done repeatedly till the parameters converge to steady-state values.

5.4.3.1 Overview of the Model

First, the MBMAP algorithm is modeled to evaluate the probability of a burst having n packets which also gives the burst size distribution. Next, the TCP source is modeled as a Markov chain with different states representing the phases in the evolution of congestion window. The collection of packets from a single source in a burst called *per-flow-burst* (PFB) is used to enable the exact evaluation of the packet loss probability from the BLP. With this model the number of packets sent in each state is computed which helps to estimate the impact of a burst loss on a TCP flow. The end-to-end delay is evaluated using a queueing theoretic model for the core network. The average BLP and the delay obtained from the network model are used by the source model to compute the throughput of a sender. A fixed point iteration method is used to alternately solve the model for TCP sender and the OBS network to characterize the behavior of TCP in OBS networks with good accuracy and provide several insights.

5.4.3.2 Modeling the MBMAP Assembly Algorithm

Figure 5.10 shows the topology with the TCP connections at the edge node used in this model. Multiple TCP flows are established between the ingress nodes (nodes numbered 1–4) and the egress nodes (nodes numbered 7–10) in which the through-

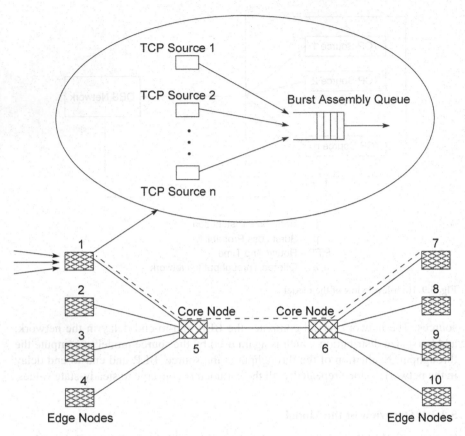

Fig. 5.10 Network used in the model

put of a flow between 1 and 7 (shown in dotted line) is studied. On the path shown from node 1 to node 7, there is an ingress node wherein packets from several TCP flows destined to the same egress node (n flows shown in figure) are aggregated in a burst.

The MBMAP assembly algorithm [7] generates a burst when the burst size exceeds the burst length threshold (Bth) or when the assembly period (T_b) is exceeded, whichever happens first. This assembly algorithm is more generic compared to the time-based or size-based assembly and does not penalize slow TCP flows [7]. Let E_{BAT} be the event that the assembly time T_b is exceeded and E_{BTH} be the event that the burst size exceeds Bth. Assuming that there are n packets in a burst, the E_{BAT} event (assembly through expiry of the timer) occurs when the sum of inter-arrival time of $n-1$ packets is less than T_b, whereas the sum of inter-arrival time of n packets exceeds T_b. In this case the sum of size of packets should also be less than Bth (otherwise E_{BTH} would have occurred). Similarly, the event E_{BTH} occurs only when the probability that the total size of $n-1$ packets does not exceed the threshold burst size Bth whereas for n packets it does, along with the condition that the sum of inter-arrival time of n packets is less than T_b. Also the packet inter-arrival time is exponentially distributed because the packet arrival process follows a Poisson process. Since only one of these events occurs during burst assembly, the probability mass function (pmf) of the number of packets in a burst (N_p) denoted by f_{N_p} which also gives the burst size distribution is given by [2],

$$f_{N_p}(n) = \text{Prob}[N_p = n, E_{\text{BAT}} \text{ occurs}] + \text{Prob}[N_p = n, E_{\text{BTH}} \text{ occurs}]. \quad (5.63)$$

$\text{Prob}[E_{\text{BAT}}]$ and $\text{Prob}[E_{\text{BAT}}]$ can be derived by conditioning on the number of packets in the burst and by using Eq. 5.63.

5.4.3.3 Modeling the TCP Congestion Window

At each epoch of change, the window size is independent of the past values and depends only on the successful delivery or the loss of packets sent in the current window. In view of this the evolution of the window (W) can be modeled as a stochastic process using a homogeneous discrete-time Markov chain represented by a sequence $\{W_k\}$ where k is the epoch of change in the window size. The evolution of congestion window is modeled in complete detail by considering all the phases of the protocol (TCP NewReno is considered here) namely, slow start (SS), congestion avoidance (CA), fast retransmit/fast recovery (FR), time-out (TO), and exponential back-off states (B). The state space that includes all the five phases with all the possible range of values for W is defined by

$$(\text{SS}, W, W\text{th}) \in \{(\text{SS}, 1, 3), (\text{SS}, 2, 3), (\text{SS}, 1, 4), \dots, (\text{SS}, 2^{\lceil \log_2 \lceil W_m/2 \rceil \rceil - 1}, \lceil W_m/2 \rceil)\}$$
$$(\text{CA}, W) , \quad W \in \{1, 2, 3, \dots, W_m\}$$
$$(\text{FR}, W) , \quad W \in \{4, 5, \dots, W_m\}$$
$$(\text{TO}, W) , \quad W \in \{2, 3, \dots, W_m\}$$
$$(B, \text{bck}) \in \{(B, 4), (B, 8), (B, 16), (B, 32), (B, 64)\}, \quad (5.64)$$

where W_m is the maximum window size and $W\text{th}$ is the slow start threshold. The trivial case of slow start threshold $W\text{th} = 1, 2$ is built into the CA phase. The parameter bck in B states denotes the exponential back-off factor. When a loss is detected from triple duplicate ACKs, the window size at that time should be at least 4 (at least three packets have to be sent successfully after the lost packet(s)). So FR states start from a window size of 4. The case of a packet loss immediately at the start of a TCP connection (when $W = 1$) is neglected, therefore TO states start from $W = 2$. It can be seen from the definition of state space that its size depends on W_m.

To model the impact of burst loss on the evolution of window accurately, the number of packets lost from a single TCP flow in a burst loss needs to be determined. Since there might be several TCP flows to the same egress node, an assembled burst might contain packets from several flows. As stated earlier, the set of packets in a burst that belong to a single TCP source is called a PFB. Each burst can be seen as a collection of PFBs each of them from different sources. The order and the number of

PFBs in a burst depend on the arrival rate of each TCP flow and the burst formation time. Depending on the size of the burst, it can have only one PFB or many PFBs. Let *PFB* be the random variable which denotes the number of packets from a single TCP source in a burst with a distribution $f_{PFB}(n)$ given by

$$
\begin{aligned}
f_{PFB}(n) \\
= \mathrm{Prob}[PFB = n | E_{\mathrm{BAT}}]\mathrm{Prob}[E_{\mathrm{BAT}}] + \mathrm{Prob}[PFB = n | E_{\mathrm{BTH}}]\mathrm{Prob}[E_{\mathrm{BTH}}]
\end{aligned}
\tag{5.65}
$$

$$
= \left(\frac{(\lambda_i T_b)^n e^{-\lambda_i T_b}}{n!} \right) \mathrm{Prob}[E_{\mathrm{BAT}}] + \left(\lim_{t \to T_b} \int_0^t \frac{(\lambda_i t)^n e^{-\lambda_i t}}{n!} f_T(t) dt \right) \mathrm{Prob}[E_{\mathrm{BTH}}].
\tag{5.66}
$$

Equation 5.65 computes the pmf by conditioning on the two possible events in the burst assembly (E_{BAT} and E_{BTH}). When assembly happens with E_{BAT} event, *PFB* is the number of packets which arrive with a mean rate λ_i (arrival rate of the TCP flow) in the time interval T_b. For the event E_{BTH}, the number of packets that arrive during the burst formation time, t, is computed by using the law of total probability to integrate over all values of t from 0 to T_b and use the pdf of the burst formation time $f_T(.)$ (where T is a random variable denoting the burst formation time) to obtain Eq. 5.66. Note that the maximum time for the burst formation in the event E_{BTH} is T_b after which E_{BAT} occurs. $f_T(\cdot)$ is obtained by conditioning on the number of packets in a burst and using the fact that inter-arrival time of packets is exponentially distributed.

By modeling the burst as a composition of PFBs it can be seen that when a burst is lost each TCP source loses a *PFB* of packets. Hence, the probability of losing a *PFB* of packets that belong to a source is the same as the probability of a burst loss, i.e., BLP. To compute the change in the window size W of the TCP flow after a burst loss, the distribution of *PFB* (number of packets in it) given W should be evaluated. Let $f_{N_{PFB}|W}$ be the conditional pmf of the number of PFBs (denoted as N_{PFB}) in a window of size W and PFB_i be the sequence of iid random variables with the distribution of f_{PFB}. The probability that there are n PFBs in a window of size w is given by [3]

$$
f_{N_{PFB}|W}(N_{PFB} = n | W = w) = \mathrm{Prob}[\sum_{i=1}^{n} PFB_i = w].
\tag{5.67}
$$

Note that $f_{N_{PFB}|W}(n|w)$ can be obtained by an n-fold convolution of the pmf of *PFB*.

5.4.3.4 State Transition Probabilities of Markov Chain

When a burst is lost, the next phase in the congestion control depends on the *PFB* size distribution. From the model for source it can be seen that the system can exist in one of the five sets of states. Next, the transition probabilities between different sets of states are derived.

Let $P_{NL}(W)$ be the probability that none of the packets in a window W is lost, $P_{TO}(W)$ be the probability of a TCP sender timing out before receiving less than three duplicate ACKs for W packets sent, and $P_{FR}(W)$ be the probability of a TCP sender going to an FR state on receiving triple duplicate ACKs. All the transition probabilities are derived by conditioning on the number of PFBs lost in W packets with a distribution given by Eq. 5.67. Let p be the BLP which is also same as the probability of losing a PFB. Therefore the probability of successfully sending a window given that there are n PFBs in W is $(1 - p)^n$. Therefore $P_{NL}(W)$ can be written as

$$P_{NL}(W) = \sum_{n=1}^{W} (1 - p)^n f_{N_{PFB}|W}(n|W). \qquad (5.68)$$

Note that $1 \leq N_{PFB} \leq W$ (when $N_{PFB} = W$ each PFB has only one packet). All the PFBs should be successfully delivered for no packet loss to happen.

Before other transition probabilities are derived, let $P_{PFB}(i|W)$ be the probability of losing i PFBs in a window which is derived by conditioning on $f_{N_{PFB}|W}$ and then summing over the possible number of PFBs in a window (using the law of total probability). That is,

$$P_{PFB}(i|W) = \sum_{n=1}^{W} {}^n C_i \, p^i (1 - p)^{n-i} f_{N_{PFB}|W}(n|W). \qquad (5.69)$$

A TO loss occurs when less than three duplicate ACKs are received before the RTO expires. Let the probability of a loss resulting in less than three packets being successfully transmitted be denoted by $P_3(W)$. In such a case, the sum of size of x PFBs lost should be equal to either $W - 2$, $W - 1$, or W. Along with this condition, using the conditional probability $P_{PFB}(x|W)$, $P_3(W)$ can be written as

$$P_3(W) = \sum_{x=1}^{W} \sum_{k=W-2}^{W} \text{Prob} \left[\sum_{i=1}^{x} PFB_i = k \right] P_{PFB}(x|W)$$

$$= \sum_{x=1}^{W} \sum_{k=W-2}^{W} f_{N_{PFB}|W}(x|k) P_{PFB}(x|W). \qquad (5.70)$$

Therefore, $P_{TO}(W)$ can be written as

$$P_{TO}(W) = \begin{cases} (1 - P_{NL}), & 2 \leq W < 4 \\ (1 - P_{NL}(W)) \times P_3(W), & W \geq 4. \end{cases} \qquad (5.71)$$

When $W \geq 4$, TCP NewReno either transmits all the segments in a window success-fully or goes to a TO or FR state on a burst loss. Using the law of total probability $P_{FR}(W)$ can be written as

$$P_{FR}(W) = 1 - P_{NL}(W) - P_{TO}(W), \quad W \geq 4. \tag{5.72}$$

The transition probability from a state V_k to any other state V_{k+1} for all types of states is given by Eqs. 5.68, 5.71, and 5.72. When $W = 1$, which means only one packet in a PFB, the probability of sending it successfully is $1 - p$. For the SS states, i.e., V_k is of the type (SS, W, Wth), window size increases exponentially in the absence of a packet loss. When a loss occurs, transition can occur to a TO state or an FR state, depending on the number of packets lost and current size of the congestion window. Therefore,

$$V_{k+1} = \begin{cases} (SS, 2W, W^{th}), & \text{w.p } P_{NL}(W), & W < W^{th} \ \& \ 2W < W^{th} \\ (SS, 2, W^{th}), & \text{w.p } (1 - p), & W = 1 \\ (CA, W^{th}), & \text{w.p } P_{NL}(W), & 2W \geq W^{th} \\ (FR, W), & \text{w.p } P_{FR}(W), & 4 \leq W < W^{th} \\ (TO, W), & \text{w.p } (1 - P_{NL}(W)), & 2 \leq W < 4 \\ & \text{w.p } P_{TO}(W), & 4 \leq W < W^{th} \\ (B, 4), & \text{w.p } \ p, & W = 1 \end{cases} \tag{5.73}$$

where w.p means "with probability."

When $V_k = (CA, W)$, window size increases linearly without loss and can go to a TO, FR, or B state depending on the nature of the loss. The transition probabilities to the next state V_{k+1} are given by

$$V_{k+1} = \begin{cases} (CA, W + 1), & \text{w.p } (1 - p), & W = 1 \\ & \text{w.p } P_{NL}(W), & 2 \leq W \leq W_m \\ (CA, W), & \text{w.p } P_{NL}(W), & W = W_M \\ (FR, W), & \text{w.p } P_{FR}(W), & 4 \leq W \leq W_m \\ (TO, W), & \text{w.p } (1 - P_{NL}(W)), & 2 \leq W < 4 \\ & \text{w.p } P_{TO}(W), & 4 \leq W \leq W_m \\ (B, 4), & \text{w.p } p, & W = 1. \end{cases} \tag{5.74}$$

On receiving triple duplicate ACKs, TCP source reaches an FR state and sets $W^{th} = \max(\lceil W/2 \rceil, 2)$ when loss occurred at window size W. From an FR state, i.e., $V_k = (FR, W)$, after the fast retransmit/fast recovery phase of NewReno with partial window deflation, W is reduced by half [15]. Therefore,

$$V_{k+1} = (CA, \lceil W/2 \rceil), \text{ w.p } 1, \ 4 \leq W \leq W_m. \tag{5.75}$$

After a time-out loss, W is reset to 1 and Wth is set to $\lceil W/2 \rceil$. So when V_k is of the type (TO, W), the transition probability is given by

$$V_{k+1} = (SS, 1, \lceil W/2 \rceil), \text{ w.p } 1, \quad 2 \le W \le W_m. \tag{5.76}$$

Lastly for the BO states the transition probabilities are given by

$$V_{k+1} = \begin{cases} (CA, 2), & \text{w.p } (1-p) \\ (B, 2 * bck), & \text{w.p } p, \quad bck \ne 64 \\ (B, bck), & \text{w.p } p, \quad bck = 64. \end{cases} \tag{5.77}$$

The transition probability for all transitions other than those defined above is zero. The above-defined transition probabilities are used to define the transition probability matrix in its entirety.

5.4.3.5 Throughput of a TCP Source

From the definition of the state space and transitions between various states, it can be seen that the discrete-time Markov chain is irreducible and aperiodic. Hence, it has a stationary distribution π so that the throughput of the TCP source is given by

$$\text{Throughput} = \frac{\sum_{i=1}^{T_s} \pi(i)N(i)}{\sum_{j=1}^{T_s} \pi(j)S(j)}, \tag{5.78}$$

where T_s is the total number of states in the Markov chain, $\pi(i)$ is the stationary probability of the ith state which can be computed using any one of the numerical techniques to solve Markov chains [38], and $N(i)$ and $S(i)$ are the number of packets sent and sojourn time in the ith state of the Markov chain, respectively. To determine the throughput of a TCP source, the number of packets sent in each state and the sojourn time in each state of the Markov chain need to be evaluated. Note that the number of packets sent and the waiting time in FR states are different for each TCP variant and can be calculated on similar lines based on the fast recovery phase. The remaining expressions are the same for any loss-based variant of TCP.

5.4.3.6 Modeling the OBS Network

Next, the OBS network is modeled as a network of queues so that the average delay and the BLP in the network are evaluated which directly affect the throughput of the TCP. Evaluation of the BLP helps to compute the number of packets lost from a source which determines various transitions in the Markov chain of the source model. On the other hand, evaluation of delay at the assembler, disassembler, and core nodes is necessary to compute the RTT accurately. The theory of bulk queues is used to compute the delay at the burst assembler and disassembler and the path along the core nodes is modeled as an open queueing network with rejection blocking [4] to compute the BLP. The reduction in the load at a node due to blocking at the upstream nodes similar to that in the reduced-load fixed point approximation model presented in Section 3.8 is also considered.

The burst assembler can be modeled as an $M/M^{[K]}/1$ batch service queue to analyze the delay incurred due to assembly. For MBMAP assembly, K can be approximated with the mean burst size which can be obtained from Eq. 5.63 as $K = \sum_{n=1}^{\infty} n.f_{N_p}(n)$. This is an approximation used to estimate delay in assembly and avoid the complex case of assuming K as a random variable defined by the burst size distribution. On solving the $M/M^{[K]}/1$ system the total delay (queueing time(D_{qI}) + service time) in the queue denoted by D_I can be obtained. Similarly, to compute the delay due to disassembly at the egress, a bulk arrival queue ($M^{[Y]}/M/1$) model can be used. Y is also a random variable whose distribution is the burst size distribution given by Eq. 5.63. The total delay (D_E) and queueing delay (D_{qE}) at the disassembler can be evaluated using the theory of ($M^{[Y]}/M/1$) queue. Finally, the core network including the switches at the edge nodes can be modeled as an open network of $M/M/W_k/W_k$ queues with rejection blocking, where W_k is the number of wavelengths available at a node k [4]. An iterative algorithm given in [4] is used to calculate the net load offered and the blocking probability at each node in the network. This is used to get the average blocking probability in the network. The total delay (D_C) in the core network consists of transmission delay and propagation delay as there is no queueing delay.

Finally, the end-to-end delay for a TCP segment is computed using the delay computed at the assembler, core network, and disassembler. Therefore, the average RTT is given by RTT $= D_I + D_C + D_E$. The average retransmission time-out (RTO) is computed as RTO $=$ RTT $+$ 4RTTVAR, where RTTVAR is the standard deviation in RTT. RTTVAR can be approximated with the average queueing delay in the network (sum of queueing delay at the ingress and egress) which is given by $D_{qI} + D_{qE}$.

5.4.3.7   Fixed Point Analysis of TCP over OBS

The TCP throughput mainly depends on the packet loss probability. A high packet loss rate reduces the throughput, which in turn reduces the network load and thus the packet loss rate. The coupled relationship between the throughput and the packet loss probability is modeled using the fixed point analysis. In the Markov chain-based model presented above, the network and the source are modeled separately. However, the throughput of the source depends on the loss probability which can be obtained by solving the network model. The fixed point analysis technique helps to model the interaction of TCP flows with the components of an OBS network [3, 6].

The variation of TCP throughput with network parameters such as the RTT, the number of wavelengths, and the burst size distribution is studied using the fixed point analysis. A theoretical model for the OBS network evaluates the packet loss rate in the network based on the input load and also a model for the TCP source. The throughput is estimated as a function of packet loss rate which is then required to find the operating point of the network using fixed point analysis. Consider a bottleneck link in which N_s TCP connections are passing destined to M_s different destinations. All the packets destined for the same node are buffered together and hence there are M_s such buffers at the ingress node. The packet arrival process is assumed to be Poisson for each TCP flow with rate λ_p and service rate of μ_p. A burst is generated every T units of time and transferred to the scheduler for possible transmission. The burst arrival rate λ_B is a function of the burst formation time T and the burst service rate is μ_B both of which are given by

$$\lambda_B = \frac{1}{T},$$ (5.79)

$$\mu_B = \frac{\mu_p}{N_s T \lambda_p}.$$ (5.80)

An ON/OFF traffic source is assumed and using a two-dimensional Markov chain model, blocking probability is estimated for this topology. The TCP throughput is evaluated as a function of loss rate using Eq. 5.19. Figure 5.11 shows the four main steps involved in determining the operating point of the network: setting the TCP input rates, computing the OBS parameters based on the input rate, calculating the loss rate in the OBS network, calculating the TCP throughput based on this observed loss rate and then using this throughput again as the next input rate.

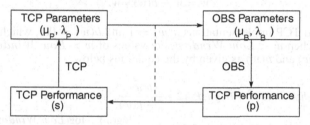

Fig. 5.11 Fixed point model

5.4.4 Modeling High-Speed TCP

HSTCP has been proposed to obtain high throughput in high bandwidth-delay product networks with low error rates. Normal TCP implementations in the congestion avoidance phase take a large number of RTTs to utilize the available bandwidth in high-speed networks. This leads to a low utilization of such links. For example, to achieve a steady-state throughput of 10 Gbps, a standard TCP connection with 1500 byte packets and RTT of 100 ms requires an average congestion window size of 83, 333 segments and a packet drop rate of at most one drop per 5 billion packets, far below what can be achieved even with the current optical network technology [16]. Random burst losses in OBS networks which could lead to a loss of multiple TCP segments at the same time degrade the throughput. HSTCP improves the performance in such situations by using a large congestion window and the additive increase and multiplicative decrease parameters are designed as functions of the congestion window instead of being constants in the TCP. Due to the aggressive nature of window growth in HSTCP, it is found to have better throughput over OBS networks compared to any variant of TCP particularly when the BLP is high.

HSTCP always has a higher average throughput than TCP irrespective of the access bandwidth, burst size, path length, and the number of flows [39].

Three parameters used in the description of HSTCP response function are Low_Window, $High\_Window$, and $High\_P$. When the current congestion window is lower than Low_Window, HSTCP has the same congestion control mechanism as TCP. Similarly, the upper bound on the HSTCP response is fixed using the parameter $High\_Window$, which is also the value of $High\_P$. The congestion window sizing mechanism in HSTCP basically follows the AIMD scheme with the increase and decrease parameters being functions of the current window size, denoted by $a(w)$ and $b(w)$, respectively. In response to an ACK packet, HSTCP increases its congestion window as

$$w = w + \frac{a(w)}{w} \tag{5.81}$$

and decreases the congestion window in response to a congestion event (TD loss) as

$$w = w - b(w) * w. \tag{5.82}$$

In standard TCP implementations $a(w) = 1$ and $b(w) = 0.5$, which is also true for HSTCP when $w < Low\_Window$. For values of $w > Low\_Window$, HSTCP calculates $a(w)$ and $b(w)$ as given by the equations below.

$$a(w) = High\_Window^2 * High\_P * 2 * \frac{b(w)}{2 - b(w)}, \tag{5.83}$$

$$b(w) = 0.5 + (High\_Decrease - 0.5) * \frac{\log(w) - \log(Low\_Window)}{\log(High\_Window) - \log(Low\_Window)}, \tag{5.84}$$

where $High\_Decrease$, Low_Window, and $High\_Window$ are predefined values [16].

For modeling HSTCP over OBS consider the simple topology shown in Fig. 5.2. Let B_a be the access bandwidth, p_b be the BLP, and T_b be the burst formation time. Since TCP sends packets based on the ACKs, the TCP packet transmission pattern is assumed to be a packet train as shown in Fig. 5.12. In this figure time is divided into RTT slots and each RTT slot contains a packet train. The length of a packet train depends on the window size. When the link capacity B_a is completely used by the TCP flow, the packet train covers the entire RTT slot. Let the length of the packet train when the window size is w in an RTT slot be denoted by $L(w)$ such that [51]

$$L(w) = \frac{w}{B_a}. \tag{5.85}$$

Since the burst assembly time is T_b, the packets sent in one RTT appear in $\lceil \frac{L(w)}{T_b} \rceil$ bursts. In one RTT, the probability of the TCP packets being lost when a burst is lost

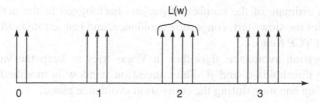

Fig. 5.12 TCP packet train

is given by

$$q(w) = 1 - (1 - p_b)^{\lceil \frac{L(w)}{T_b} \rceil}. \tag{5.86}$$

Using the fluid framework to model the TCP, a differential equation that describes the rate of change in the window size can be written as [23, 51]

$$\frac{dw}{dt} = (1 - q(w))\frac{a(w)}{\text{RTT}} - q(w)\frac{b(w)w}{\text{RTT}}. \tag{5.87}$$

The first term on the right-hand side of Eq. 5.87 represents the window increment process and the second term indicates the window decrement process. In order to obtain the window size at the equilibrium point w^*, set the derivate in Eq. 5.87 to zero.

$$(1 - q(w^*))a(w^*) - q(w^*)b(w^*)w^* = 0. \tag{5.88}$$

Therefore, the throughput of a HSTCP source is defined as

$$\text{Throughput} = \frac{w^*}{\text{RTT}}. \tag{5.89}$$

5.4.5 TCP Vegas

TCP Vegas is a delay-based TCP implementation which uses the measured value of RTT to estimate the available bandwidth in the network. Vegas uses the difference between the estimated throughput and the measured throughput to get an estimate of the congestion in the network. For every round trip time and window size W, Vegas uses the most recent RTT measurement (RTT) and the minimum RTT observed so far (baseRTT) to calculate [5]

$$\text{diff} = (\frac{W}{\text{baseRTT}} - \frac{W}{\text{RTT}})\text{baseRTT}. \tag{5.90}$$

diff gives an estimate of the number of packets backlogged in the network. TCP Vegas modifies the slow start, congestion avoidance, and fast retransmit/fast recover algorithms of TCP Reno.

The congestion avoidance algorithm in Vegas tries to keep the value of diff between two thresholds α and β. The congestion window is modified according to the following equation during the congestion avoidance phase:

$$W = \begin{cases} W+1, & \text{diff} < \alpha \\ W, & \alpha \leq \text{diff} \leq \beta \\ W-1, & \text{diff} > \beta. \end{cases} \tag{5.91}$$

The slow start algorithm is also modified in TCP Vegas. The window is doubled for every alternate RTT and based on the value of diff it exits slow start if diff exceeds a threshold γ. γ is usually taken as $\frac{\alpha+\beta}{2}$.

TCP Vegas incorporates a few changes in the fast recovery phase also. A window size of 2 is used (instead of 1) at initialization and after a time-out. When multiple packets are lost in a single window, Vegas decreases the window only for the first loss. The window size is reduced by 25% instead of 50% in such cases.

In this section, a renewal theory-based model for the throughput of TCP Vegas over OBS networks is presented which is derived on similar lines as that for the other loss-based variants [36]. The access bandwidth is considered high enough such that the entire congestion window is contained in a burst. In such a scenario, loss indications via triple duplicate ACKs are not considered. Therefore, the throughput of TCP Vegas is given by Eq. 5.26.

The IP access network is assumed not to be congested and the burst assembly delay is neglected. Hence RTT would be very close to baseRTT. Figure 5.13 shows the evolution of Vegas congestion window over OBS networks.

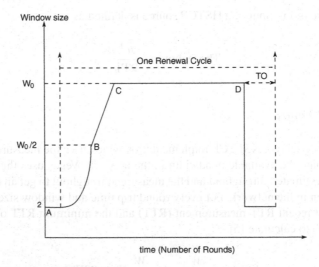

Fig. 5.13 TCP Vegas window evolution

The evolution of the congestion window is partitioned into different periods as given below:

- Slow start period which is from A to B wherein the window starts at 2 and doubles every other RTT and increases until it reaches the slow start threshold.
- Transition period from B to C, where the window increases linearly because diff is less than α as RTT is close to baseRTT. This goes on until the stable state is reached where $\alpha \leq \text{diff} \leq \beta$.
- The loss free period from C to D where no losses occur and Vegas is in the stable state.
- The time-out period which may consist of a sequence of consecutive time-out events.

In order to compute the steady-state throughput of a TCP Vegas source, the expected time spent and the number of packets sent in each of these different phases should be evaluated. Let W_0 be the size of the congestion window when Vegas is in stable state during period C to D. During this period the expected value of diff is given by

$$\text{diff} = W_0 \left(1 - \frac{\text{baseRTT}}{\text{RTT}} \right) = \frac{\alpha + \beta}{2} \tag{5.92}$$

and W_0 can be computed using Eq. 5.90 as

$$W_0 = \min \left(\frac{\alpha + \beta}{2} \times \frac{\text{RTT}}{\text{RTT} - \text{baseRTT}}, W_m \right). \tag{5.93}$$

During the slow start period from A to B, the congestion window doubles every other RTT until it reaches the slow start threshold $W_0/2$. Therefore, the number of packets sent during A to B is given by

$$Y_{AB} = 2 \sum_{i=0}^{\log \frac{W_0}{4}} 2^i = 2^{\log W_0} - 4 \tag{5.94}$$

and the duration of this period is

$$A_{AB} = 2 \frac{\log W_0 - 2}{\text{RTT}}. \tag{5.95}$$

During the linear increase phase of window from B to C, window increases from $W_0/2$ to W_0. The number of packets sent during this period is

$$Y_{BC} = \sum_{i=W_0/2}^{W_0-1} i. \tag{5.96}$$

As the window increases by 1 every RTT during this period, the expected duration is given by

$$A_{BC} = \frac{W_0}{2} RTT. \tag{5.97}$$

The average number of successful rounds based on Eq. 5.29 which includes the sending rounds from A to D is given by $E[X] = \frac{1-p_b}{p_b}$. Hence, the number of rounds from C to D is given by

$$S_{CD} = \begin{cases} 0, & \text{if } (\frac{1-p_b}{p_b} < 2(\log W_0 - 2) - \frac{W_0}{2}) \\ \frac{1-p_b}{p_b} - 2(\log W_0 - 2) - \frac{W_0}{2}, & \text{otherwise.} \end{cases} \tag{5.98}$$

From the above equation, the duration of stable period can be computed as

$$A_{CD} = RTT \times S_{CD} \tag{5.99}$$

and because the window size is constant during this period the expected number of packets sent is calculated as

$$Y_{CD} = W_0 S_{CD}. \tag{5.100}$$

The expected number of packets sent during the time-out phase $E[H]$ and the duration of this period $E[Z^{TO}]$ is same as explained in Section 5.4.1. Therefore, the final throughput of TCP Vegas over OBS is given by

$$B_{Vegas} = \frac{Y_{AB} + Y_{BC} + Y_{CD} + E[H]}{A_{AB} + A_{BC} + A_{CD} + E[Z^{TO}]}. \tag{5.101}$$

Based on the expression for the throughput obtained above, the performance of TCP Vegas over OBS networks is evaluated with burst-level retransmission and deflection routing. When a fixed source routing is used, the packet delay experienced in the OBS network is mainly due to the burst assembly apart from the propagation delay along the path. These factors do not vary with traffic load so that the conventional Vegas cannot detect congestion effectively. Therefore, TCP Vegas detects congestion falsely which impairs the performance significantly. Based on these observations, a threshold value is used for the RTT to differentiate between congestion and burst losses at low traffic loads. The sender measures the statistics of RTT values over the time which is used to distinguish false congestion [37].

5.5 TCP Implementations in OBS

A comparison of the performance of TCP Reno, NewReno, and SACK over OBS shows that the throughput falls drastically due to the burst losses which cause simultaneous loss of several packets [46]. For medium and slow rate TCP flows, when the number of TCP segments in a burst is small, a burst loss is equivalent to multiple TD losses. But for fast TCP flows, where the assembly time at the ingress is sufficiently large to make a burst out of an entire window of packets, a burst loss usually leads to a time-out. Even with a single burst loss, the throughput of all the three implementations falls drastically. NewReno performs worse than Reno when the length of the lost burst is large because NewReno retransmits only one packet in each RTT and does not send any new packets as long as all the lost packets are not retransmitted. This causes NewReno to take longer time than Reno to complete the retransmission of all the lost packets. SACK has the best performance out of all the three variants because, the ACK contains explicit information on the number of packets lost and all the packets are retransmitted in one RTT. Further, SACK halves the congestion window only once for all the lost packets. For a fast TCP flow the performance of all the implementations is the same since, a burst mostly contains all the packets sent in one round. In presence of multiple burst losses, all the implementations tend to have similar performance irrespective of the speed of a TCP source because, the losses lead to time-out in most of the cases. It is also observed that NewReno has better performance when the assembly time is small while Reno has better performance for larger assembly time values. In general the relative performance of the TCP implementations depends on the relationship between the RTO and the product of burst length and RTT.

5.5.1 Burst TCP

The first TCP implementation proposed for OBS networks was the burst TCP (BTCP) which showed that differentiating between packet losses due to congestion and packets lost due to a single burst loss improves the performance of TCP over OBS [46]. BTCP checks if a TO event is due to the loss of multiple packets in either one of the access networks. If it is so then it is a true TO. Else if multiple packets are lost because of a single burst loss in OBS network it is an FTO. BTCP also considers packet losses due to multiple burst losses in the OBS network as a true TO. BTCP includes three techniques to detect the FTOs and treat them as TD losses which is actually the way they are supposed to be treated. The first technique used by the BTCP does not use explicit feedback from the OBS nodes, is simple to implement, and can improve the throughput without being affected by the FTOs. It estimates the maximum number of TCP packets in a burst sent. This method is called burst length estimation (BLE) where the BTCP sender maintains additional piece of information which is the burst window size ($burst\_wd$) to estimate the current number of packets contained in a burst. The BLE method requires no coordination or information exchange between the BTCP sender and the OBS nodes. When a TO occurs the sender compares the current size of the congestion window ($cwnd$) with $burst\_wd$ to decide whether it is a true TO or FTO. The value of $burst\_wd$ is set based on the effect of loss on the TCP sender.

Initially before the slow start phase $burst\_wd$ is set to 0. If the first loss is a TD loss, $burst\_wd$ is set to $cwnd/2$ else if it is a TO loss which is treated as an FTO and $burst\_wd$ is set to $cwnd$. If the loss is not the first one and it is a TD loss, then $burst\_wd$ is set to $\min(burst\_wd, cwnd/2)$, but if it is the first TO loss (though the loss is not the first one), it is set to $\min(burst\_wd, cwnd)$, else it is set to $\max(burst\_wd, cwnd)$. Once the value of $burst\_wd$ is initialized according to the type of the loss, it is simply compared with $cwnd$ and decision is made if the TO is an FTO or not. For any TO loss, as long as $cwnd \leq burst\_wd$ and $burst\_wd > 3$, the sender treats the TO as an FTO and halves its congestion window and starts fast retransmission of the packets lost. Otherwise for a TO loss with $cwnd > burst\_wd$ and $burst\_wd \leq 3$, the sender identifies the TO similar to the other TCP implementations.

The burst window estimated by the BLE method is not necessarily accurate and cannot distinguish between multiple packet losses in the access network from a single burst loss in the OBS network. It might identify a true TO with an FTO for medium rate flows. The second technique used by the BTCP which is more accurate than the BLE method requires the ingress node to send a burst ACK (BACK) that contains the information of the TCP packets assembled in a burst. This way a sender knows if all the packets lost are contained in a single burst or not. On a TO loss, a sender uses the BACK to check if all the packets lost are sent in a single burst. If this is the case, it is treated as an FTO. However, the sender has no way of knowing if the packets are lost in the OBS network or the access network in the receiver side. If the BTCP sender did not receive a BACK before the timer expires, it is assumed to be a true TO caused by the loss of BACK or the packets sent earlier in the sender-side access network. To identify the loss of packets in the receiver-side access network and thus increase the accuracy of identifying FTOs a BACK can be sent both by the ingress and by the egress nodes. BTCP with BACK can achieve better throughput than the other TCP implementations by identifying true TOs accurately. Nevertheless the OBS edge nodes are required to process the TCP packets to send the BACK.

The third scheme of BTCP, which is BTCP with burst negative ACK (BNAK), requires each control packet to contain the information on TCP packets within the corresponding burst. When a burst is lost, the core node constructs a BNAK packet using the information in the control packet to inform the BTCP sender the exact packets lost. BNAK-based BTCP implementation achieves the same accuracy of BTCP with BACK from both the edge nodes. However, it has lower RTT since the lost packets can be retransmitted faster. BTCP sender with BNAK need not wait for the TO to occur for each packet and can start retransmission of packets as soon as BNAK is received.

Based on the theoretical model proposed for the throughput of TCP in OBS networks described in Section 5.4.1 a simple expression for the throughput of BTCP with all the three techniques of detecting FTOs is derived here.

- BTCP using BLE or BACK: Let p be the BLP assumed to be an iid random variable. The average number of rounds successfully transmitted between two

TO losses is $\frac{1-p}{p}$ (see Eq. 5.29). When the loss rate is not very low and the maximum window is relatively large such that $W_m > 1/p$, if TO losses are distributed uniformly over time, the sender starts retransmission from W_0 and reaches $W < W_m$ before the next TO loss. For a fast TCP flow (Reno, NewReno, or SACK), $W_0 = 1$ and $E[W] = \frac{1-p}{p}$ and during the slow start phase, the TCP sender transmits $W - 1$ packets in $\log_2(\frac{E[W]}{2})$ rounds and $W/2$ rounds during the congestion avoidance phase. Therefore the average number of packets transmitted between two TOs denoted by $E[N_1]$ is written as

$$E[N_1] = (E[W] - 1) + \frac{E[W]}{2} \times (\frac{E[W]}{2} + E[W])/2 = \frac{3}{8}E[W]^2 + E[W] - 1.$$
(5.102)

By equating the expressions for the average number of rounds transmitted between two TO events

$$\frac{1-p}{p} = \log_2(\frac{E[W]}{2}) + \frac{E[W]}{2}.$$
(5.103)

$E[W]$ can be obtained which then gives $E[N_1]$. For a fast TCP flow, the total number of packets transmitted between two TO events is given by

$$E[N_2] = \frac{1-p}{p} \times (\frac{E[W + \frac{W}{2}]}{2}) = \frac{3(1-p)^2}{2p^2}.$$
(5.104)

If the access network is assumed to be lossless the expression for the throughput is same for BTCP with BACK mechanisms. Further, the time period between two TO events (called the TOP) in BTCP with BLE/BACK mechanism is the same as that in TCP without FTO detection. Therefore, the ratio of the throughput of BTCP to that of TCP without FTO detection is simply $E[N_2]/E[N_1]$ which is obtained from Eqs. 5.102 and 5.104.

When $W_m < 1/p$, congestion window reaches W_m most of the time. In such a case, $E[N_1]$ and $E[N_2]$ are given by

$$E[N_1] = \frac{W_m}{p} - \frac{W_m^2}{8} + W_m - 1 - W_m \log_2(\frac{W_m}{2}),$$
(5.105)

$$E[N_2] = (\frac{1}{p} - \frac{W_m}{2}) \times W_m + \frac{3W_m^2}{8} = \frac{W_m}{p} - \frac{W_m^2}{8}.$$
(5.106)

Thus, the ratio of the throughputs is given by

$$\frac{E[N_2]}{E[N_1]} = 1 + \frac{W_m(\log_2 \frac{W_m}{2} - 1) + 1}{E[N_1]}.$$
(5.107)

From the equation above, it can be seen that the benefit of FTO detection in BTCP increases with the maximum window size. This is obvious because BTCP can start with a larger window size, i.e., $W_m/2$ after each FTO while the TCP implementation without FTO detection has to start from $W = 1$. Higher loss rate also increases the benefit of FTO detection because there are more TO events and detecting FTOs is essential.

- BTCP with BNAK: For a BTCP sender using BNAK mechanism, the number of packets transmitted between two TO events is the same as that by a TCP sender without FTO detection. When an FTO occurs, retransmission of packets can start as soon as a BNAK is received which can be approximated to the RTT after the lossy round (round in which packets are dropped) instead of the conventional RTO time. Therefore, the TOP in BTCP with BNAK is given by $\text{TOP}_2 = \frac{\text{RTT}}{p}$ while for a TCP sender without FTO detection it is $\text{TOP}_1 = \frac{1-p}{p}\text{RTT} + \text{RTO}$. The throughput enhancement with BTCP using BNAK mechanism can be written as

$$\frac{E[N_2]/\text{TOP}_2}{E[N_1]/\text{TOP}_1} = \frac{E[N_2]}{E[N_1]}\left(1 + \frac{\text{RTO} - \text{RTT}}{\text{RTT}} \times \frac{p}{1-p}\right). \qquad (5.108)$$

From the above equation it can be observed that BTCP using BNAK has better performance than BTCP with BLE/BACK and this improvement is greater with increasing p.

Note that the analysis does not consider the case of consecutive FTOs which are common at high p and uneven loss distribution.

5.5.2 Generalized AIMD

The next three TCP implementations proposed for OBS networks are based on the generalized AIMD (GAIMD) rate control mechanism [43]. In this scheme the factor α used to increase the window size on a successful packet delivery and the factor β used to decrease the window size for a TD loss are varied according to the nature of the loss. Most of the TCP implementations follow the AIMD scheme where α and β are set to 1 and 0.5, respectively. Such AIMD-based implementations suffer a major setback when used over OBS network because burst losses need not be caused by congestion alone. For losses that can be classified as FTOs the sender either goes to a TO or repeatedly decreases the window by 0.5 (which is β in AIMD) and might result in a small window size most of the time and thus low throughput. GAIMD has the framework to set any value to (α, β) pair according to the congestion status of the path. The main disadvantage found with BTCP is that it requires feedback from the core nodes or edge nodes and needs the OBS nodes to understand TCP packets. This causes signaling overhead and may not be practical. To overcome these disadvantages, GAIMD uses rate control at the edge node (where the TCP sender is located) to adaptively increase or decrease the throughput of TCP and maintain friendliness with the other flows.

5.5.2.1 BAIMD

The first implementation discussed is called burst AIMD (BAIMD) which uses a congestion parameter $0 < p < 0.6$ to differentiate congestion from contention losses [30]. The probability of the bottleneck link being full computed from the system of equations for a birth–death Markov process is mapped to p which is used to compute β as $\beta = 1 - p$. Once β is computed, α is computed from the Eq. 5.109 for each type of loss. For the state of congestion $p > 0.6$ and thus β is set to 0.5 and BAIMD follows the AIMD TCP. At the edge node where the TCP flows originate, both BAIMD and TCP detect packet losses using either the TD losses or the TO events. For a TD loss, the TCP sender reduces the window by a factor of 0.5 while the BAIMD sender reduces by a factor $0.5 < \beta < 1$. In the state of congestion, which is determined by measuring the RTT of the packets at the ingress node, BAIMD sender also uses $\beta = 0.5$. Similarly, for each RTT the increase in the window size is given by α which is set to 1 in TCP, whereas in BAIMD it is set according to the formula given by GAIMD scheme [30, 43]

$$\alpha = \begin{cases} \frac{3(1-\beta)}{1+\beta}, & \text{if the loss is a TD loss} \\ \frac{4}{3}(1 - \beta^2), & \text{if the loss is a TO loss.} \end{cases} \tag{5.109}$$

One of the main advantages of BAIMD is that it is simple to implement and assumes no explicit notifications from OBS nodes nor does it require to know the number of packets in a burst. However, the identification of congestion from RTT of individual packets is not easy. RTT of a packet does not determine if the path is congested or not unlike the case of packet switched networks where there is queueing delay at the intermediate nodes.

5.5.2.2 TCP-ENG

BAIMD cannot distinguish between the loss of several packets due to single burst loss and that due to multiple burst losses. With BAIMD, every burst loss is identified as that due to congestion. The estimation of congestion parameter which is used to determine the decrease factor is also inaccurate in many cases. To avoid the reduction of cwnd even when the loss is not due to congestion, explicit loss notification (ELN) is sent from the OBS ingress node [34, 31]. ELN notifies the TCP sender about the exact condition of the path (congested or not) so that the values of α and β can be decided. Every ingress node measures the utilization of the path by measuring the BLP in a time window. If the BLP is below a predefined threshold when a burst loss occurs, ELN packet is sent from the ingress to indicate that the loss is not associated with congestion. If the TCP sender does not receive any ELN packet and $1 < \text{cwnd} < 1/(1 - \beta)$, α is set to 1. If an ELN packet is received along with the packet loss, α is set according to Eq. 5.109. The slow start phase remains the same as in the AIMD scheme and only during the congestion avoidance phase, the increase factor is decided based on the information given by the ELN packet. This scheme which incorporates the advantages of both GAIMD scheme and the

ELN scheme is called TCP explicit notification GAIMD (TCP-ENG) [34]. TCP-ENG uses ELN packets to indicate that the bursts dropped are not due to congestion (which is identified with the path utilization) and thereby prevents reduction in α and increases the throughput. This scheme is more scalable than the BTCP since it does not use notifications for every loss observed. It improves the performance of TCP compared to BAIMD by identifying congestion with better accuracy (uses explicit notifications to indicate congestion).

The throughput of a TCP-ENG sender can be derived using the expression for the throughput of a TCP SACK sender given by Eq. 5.12.

Each of the terms in the expression are computed similar to the analysis used in Section 5.4.2. $E[Y]$ is given by

$$E[Y] = (\beta + 1)E[W_X] + \frac{1-p}{p}S. \tag{5.110}$$

For high burst losses, i.e., when $W_X < W_m$,

$$E[Y] = \frac{\alpha\beta - \frac{b(\beta-1)}{2} + \sqrt{(\frac{b(\beta-1)}{2} - \alpha\beta)^2 + \frac{2\alpha Sb(1-\beta^2)}{p}}}{b(1-\beta)} + \frac{1-p}{p}S, \tag{5.111}$$

$$E[X] = \frac{\alpha\beta - \frac{b(\beta-1)}{2} + \sqrt{(\frac{b(\beta-1)}{2} - \alpha\beta)^2 + \frac{2\alpha Sb(1-\beta^2)}{p}}}{1+\beta}, \tag{5.112}$$

$$E[TDP] = \overline{RTT}\left(\frac{2\alpha\beta - \frac{b(\beta-1)}{2} + \sqrt{(\frac{b(\beta-1)}{2} - \alpha\beta)^2 + \frac{2\alpha Sb(1-\beta^2)}{p}}}{1+\beta}\right). \tag{5.113}$$

For low burst losses, i.e., when $W_X = W_m$, $E[W_x] = W_m$ so that

$$E[Y] = (\beta + 1)W_m + \frac{1-p}{p}S, \tag{5.114}$$

$$E[TDP] = \overline{RTT}\left(\frac{W_m(1-\beta)}{\alpha} - \frac{W_m(1-2\beta+\beta^2)}{2\alpha} + \frac{\beta}{\alpha} + \frac{S}{\alpha W_m p} + 1\right). \tag{5.115}$$

The behavior of TCP-ENG and TCP SACK is same for TO losses. So similar to the analysis in Section 5.4.2,

$$E[H] = \frac{p}{1-p}, \tag{5.116}$$

$$E[TOP] = \frac{RTT(f(p))}{p(1-p)}, \tag{5.117}$$

where $f(p) = 1 + p^2 + 2p^2 + 4p^3 + 8p^4 + 16p^5 + 32p^6$ and $Q(E[W_X]) = p^{\frac{W_X}{S}-1}$.

Substituting Eq. 5.110 through Eq. 5.117 in Eq. 5.12 gives the throughput of TCP-ENG for the cases of both low and high losses.

5.5.2.3 SAIMD

The third scheme that uses the GAIMD framework is called statistical AIMD (SAIMD) in which a TCP sender statistically analyzes a certain number of previous RTTs and dynamically adjusts the values of α and β [33, 35]. It does not use any explicit feedback from the OBS nodes and requires only RTTs measured at the sender. The idea behind SAIMD is that the statistics of RTTs (mean, standard deviation, and autocorrelation) measured over a short time when compared with the long-term RTTs can identify a situation of congestion accurately. It is assumed that RTT of a burst increases in the core network as well as at the edge nodes when the path is congested due to delay in the scheduling and the delay at the FDLs. Comparing the average of N short-term values of RTT with M values taken over much longer period of time determines the likelihood of a packet loss due to congestion assuming that majority of the losses are due to random contentions. To determine the value of N, autocorrelation of N RTTs is calculated and a value of N that maximizes the autocorrelation is chosen.

Once the value of N is selected carefully, the next step is to map the value of average RTT (over N values) to β. It is assumed that the RTTs can be modeled using a Gaussian distribution and a function $z_i = \text{rtt}(u_i)$ which returns an RTT value larger than a u_i ($0 < u_i \leq 1$) portion of values in the distribution of RTTs where u_i is the confidence level. When average RTT is smaller than $z_1 = \text{rtt}(u_1)$, a low confidence of congestion is indicated so that β is set to 1 and there is no difference in the $cwnd$ after a loss. When average RTT is larger than $z_n = \text{rtt}(u_n)$ there is an indication that the network is congested so that $\beta = 0.5$. For all other values of average RTT in the interval $[z_1, z_n]$, $\beta = f(u_i) = 1 - \frac{u_i - u_1}{2(u_n - u_1)}$. The two parameters u_1 and u_n are set to 50 and 90%, respectively, to distinguish congestion from contention losses. The SAIMD scheme is particularly suitable for long-lived TCP flows with large $cwnd$. In such cases, the time taken to reach the window size at the time of the loss again after an FTO can be very long. The only extra cost incurred in SAIMD compared to other TCP implementations is the maintenance of M and N RTTs and computation of autocorrelation and confidence intervals for the N RTTs.

The analysis of SAIMD is exactly similar to that of TCP-ENG except that in SAIMD, α is always set to 1 and β is a random variable with Gaussian distribution function on RTTs. Thus, instead of using β in the analysis for TCP-ENG, $\overline{\beta}$ which is $E[\beta]$ given by [35]

$$E[\beta] = \overline{\beta} = \sum_i \beta_i p_\beta(\beta_i) \tag{5.118}$$

is used where $p_\beta(\beta_i)$ is the probability that β_i has distinct values.

Unlike TCP-ENG, SAIMD does not use explicit notifications due to which the packets are retransmitted only at the end of TO period. Thus,

$$E[\text{TOP}] = \text{RTO} \frac{f(p)}{1 - p}. \tag{5.119}$$

In all the other terms of Eq. 5.12, by replacing all the occurrences of α in the analysis of TCP-ENG with 1 and β with $\overline{\beta}$ gives the throughput of the SAIMD.

5.6 Improving the Performance of TCP

As mentioned in earlier sections, the performance of TCP is affected by the losses and the burst assembly algorithm in OBS networks. Several techniques are proposed to improve the performance of TCP over OBS networks which either use contention resolution techniques at the core nodes to reduce the BLP or use explicit feedback on the nature of the loss to avoid the reduction of congestion window for FTOs. The TCP variants discussed in the previous section adapt the congestion control to consider the two types of losses in OBS networks, i.e., congestion and contention. In this section, a few of the techniques used to improve the performance of TCP over OBS networks are presented. It was found that designing burst assembly algorithm to reduce the BLP and loss rate simultaneously also helps to improve the performance of TCP.

It was found that using deflection routing for contention resolution affects the performance of the TCP [29]. Deflection routing causes reordering of TCP packets at the receiver because some bursts that carry packets of the same TCP flow are routed on a longer path. Reordered packets are treated as lost packets at the receiver so that the TCP throughput is reduced due to the deflection routing. However, it was found that the impact of reordering of packets due to deflection routing is overcome by the benefit of deflection routing in reducing the loss rate. Therefore, deflection routing improves the performance of TCP as long as the difference in the delay on different paths is not significantly large. With increasing number of packets of the same TCP flow in a burst, the improvement due to deflection routing increases specially with fast TCP flows.

Using FDLs at the core nodes reduces the contention losses but increases the delay encountered by each burst. Exact analysis of the impact of FDLs requires the evaluation of the increased delay and reduced BLP on the throughput [19]. Apart from the increased delay, FDLs also reduce the BLP which improves the TCP throughput. Exact analysis of the impact of FDLs requires the evaluation of the increased delay and reduced BLP on the throughput [19].

5.6.1 Adaptive Burst Assembly Algorithm

The adaptive assembly period (AAP) algorithm adjusts the assembly period dynamically at each ingress node according to the length of the burst recently sent [7, 28]. Let q_{sd} be a queue corresponding to an ingress–egress pair and traffic from a class of service, and Q_{ij} be the set of all queues that generate traffic passing through a link

(i, j). Let T_s be the assembly period used at the node s. The average burst length at a queue, $\text{BL}_{q_{sd}}$, satisfies the inequality

$$\sum_{q_{sd} \in Q_{ij}} \frac{\text{BL}_{q_{sd}}}{T_s} \leq WB, \tag{5.120}$$

where W is the number of wavelengths and B is the capacity of each wavelength on the outgoing link. To prevent a TCP source from timing out, the delay at the ingress node should not be larger than RTO − RTT so that the assembly period at a node s (neglecting the delay in the transmission queue) is constrained by

$$\frac{\sum_{q_{sd} \in Q_{ij}} \text{BL}_{q_{sd}}}{WB} \leq T_s \leq \min_f(\text{RTO} - \text{RTT}), \tag{5.121}$$

where f denotes the set of all flows from s to d. Writing similar inequality for the assembly period at an individual queue q_{sd} gives

$$T_{q_{sd}} = \alpha \times \frac{\text{BL}_{q_{sd}}}{WB}, \tag{5.122}$$

where $\alpha \geq 1$ is called the assembly factor which indicates the average number of assembly queues sharing a link. From Eqs. 5.120 and 5.122 this factor can be written as

$$\alpha = \frac{WB}{\frac{\text{BL}_{q_{sd}}}{T_{q_{sd}}}} \simeq |Q_{ij}|, \tag{5.123}$$

where $|Q_{ij}|$ is the size of Q_{ij}. The average burst length $\text{BL}_{q_{sd}}$ is then given by

$$\text{BL}_{q_{sd}} = w_1 \times \text{BL}_q + w_2 \times \text{SBL}_q \tag{5.124}$$

where BL_q is the smoothed average burst length, SBL_q is the most recently sampled average burst length, and $w_1 = 1/4$, $w_2 = 3/4$ are positive weights. Giving larger weight to recently sampled bursts synchronizes the assembly algorithm more with the TCP sender so that after a long burst is transmitted successfully, it is better to increase the assembly period because the probability of successfully sending another burst of a larger size is higher. AAP updates the statistical average burst length and thus the assembly period to satisfy Eq. 5.122. To prevent the burst length from increasing or decreasing too much, a lower and upper thresholds ($\beta > 1$ and $0 < \gamma < 1$) are used. Though AAP introduces higher complexity at the ingress node because it requires the calculation of assembly period for each burst sent, it was shown to improve the throughput of the TCP compared to the time-based assembly algorithm [7]. Adapting the burst assembly period using the learning automata to learn the network conditions from the observed BLP was also found to reduce the loss rate irrespective of the type of traffic [41]. Assembling TCP flows based on the current window size in different assembly queues also was found to improve the performance. Different TCP flows are characterized into fast, medium, and slow class flows based on the current size of congestion window and they are aggregated into bursts using different queues so that loss of bursts has different impacts on different flows. Such a burst assembly was found to improve the throughput of TCP for all rates of flows [28, 42]. When several TCP flows are aggregated at the ingress node, synchronization among the flows also degrades the throughput. Due to burst loss, if all the flows simultaneously decrease their congestion window, the throughput drastically falls down. To avoid the effect of synchronized window decrease, multiple queues are maintained for each destination node and flows are randomly assigned to each queue [18].

Due to the interplay between the delay penalty and the delayed first loss gain, it was found that there exists an optimal assembly period which maximizes the TCP throughput for a certain loss probability [42]. Another burst assembly algorithm, the dynamic assembly period (DAP) algorithm [25], tracks the variation of the current window size and updates the assembly period dynamically for the next burst assembled. Initially, the assembly period T_s is set to $2L_c/B$ where L_c is the current burst length and B is the access bandwidth. If the ratio of L_c to the previous burst length L_p is 2, then the assembly period is doubled otherwise, $T_s = L_c/B + T_i$ where T_i is the time for processing one TCP segment. When the assembly period reaches a maximum value T_{max} it is not incremented further. The correlation between the assembly period and the burst length is given by $T_s = 8.L_c.\text{MTU}/\text{MSS}.B$ where MSS and MTU are 1460 and 1500 bytes, respectively. Substituting $L_c = W_m$ the maximum window size of TCP gives the maximum value that can be used for assembly period. DAP was also found to improve the performance of TCP compared to time-based assembly algorithm [25].

5.6.2 Using Burst Acknowledgments and Retransmission

A simple way to improve the performance of TCP traffic is to use burst-level acknowledgments and retransmission. Since JET protocol in OBS uses one-way reservation mechanism, there is no reliable transport of bursts. The basic BHP does not support any acknowledgment function. A BHP format modified to introduce the acknowledgments for bursts is shown in Fig. 5.14 [14]. Apart from the source and destination node addresses, the BHP also has a sequence number which is increased whenever acknowledgments are used. Along with the acknowledgment field that has the sequence number of the last received burst, there is an n-bit *Bit vector* field that indicates the arrival status of last n bursts. The SYN field is set for TCP packets to differentiate them from the user datagram protocol (UDP) packets. Note that ACK and SYN cannot be set at the same time. For each burst the sequence number is used to provide unique identification along with the source and destination addresses. When an egress node receives a burst, it creates an ACK burst (BHP with ACK field set) which is essentially a BHP sent in the control plane with no additional

Dest Addr	Src Addr	Burst Type	Channel ID
SYN	ACK	Checksum	Burst Length
Sequence number			
ACK number			
Bit vector		Offset time	
Options Field			

Fig. 5.14 BHP ACK format

information. When the ingress node receives an ACK burst it can identify the lost bursts based on the Bit vector field.

Along with the acknowledgments that are used to identify lost bursts, retransmitting the lost bursts improves the performance of TCP by avoiding retransmission at the TCP layer. For each burst sent, a copy is stored in electronic format at the ingress node. As soon as the ingress node receives the ACK BHP, it identifies the lost bursts from the Bit vector field which are retransmitted immediately. To control the load, bursts are retransmitted only once and the buffer at the ingress node is cleared immediately. Alternately, controlled retransmission with probabilistic parameters to control the load due to retransmitted bursts can also be used to limit the load [40]. If retransmitted bursts are lost again, the recovery is left to the TCP layer [14]. Burst-level retransmission improves the TCP throughput because the time after which the packets lost are identified and retransmitted is much lower at the burst level and all the packets lost in a single burst (consecutive bursts) are sent at the same time [48].

When burst-level retransmissions are used for TCP traffic, the main issue is whether the recovery of lost packets is done by retransmission at TCP layer or OBS layer. This issue is resolved by evaluating the impact of the delay incurred in each layer due to retransmission on the throughput and comparing it with the throughput obtained without burst-level retransmission. As long as the end-to-end delay due to retransmissions is less than the RTO which prevents the onset of slow start phase, there is a definite improvement in the throughput. For a fast TCP flow, assuming RTO = 2RTT, the maximum time that a burst can wait at the ingress for being retransmitted can be set safely to RTT so that the packets in the successfully retransmitted burst are acknowledged before the RTO expires [48]. The same delay bound holds for a medium speed flow also. For a slow TCP flow, using burst retransmission would not be of great use because each burst contains only a single packet from a TCP flow.

To extend the analysis in Section 5.4.1 for the fast TCP flow when burst-level retransmission is used requires redefining the BLP. Let p_c be the probability of a burst contention and p_d be the probability that a burst is lost even after multiple retransmission attempts. The probability of a successful round in which a burst experiences contention but is successfully retransmitted is given by

$$p_{sr} = \frac{p_c - p_d}{1 - p_d}. \tag{5.125}$$

The probability of a successful round in which there is no burst contention is given by

$$p_{nc} = \frac{1 - p_c}{1 - p_d}. \tag{5.126}$$

Assuming that each retransmission takes an average time of T_p, the average number of retransmissions required for a burst to be successfully delivered is

$$E[r] = \sum_{i=1}^{\lfloor \delta/T_p \rfloor - 1} i p_c^{i-1}(1 - p_c) + \lfloor \delta/T_p \rfloor p_c^{(\lfloor \delta/T_p \rfloor - 1)}. \tag{5.127}$$

Hence, the average RTT experienced by a successfully retransmitted burst is $RTT_r = RTT + E[r]T_p$ so that

$$E[A] = p_{sr}E[X]RTT_r + p_{nc}E[X]RTT \tag{5.128}$$

and

$$E[Z^{TO}] = RTO\frac{f(p_d)}{1 - p_d}, \tag{5.129}$$

where $f(p_d) = 1 + p_d + 2p_d^2 + 4p_d^3 + 8p_d^4 + 16p_d^5 + 32p_d^6$, $E[H] = p_d/(1 - p_d)$, $E[X] = (1 - p_d)/p_d$ and

$$E[Y] = \begin{cases} \frac{1}{p_d^2}, & p_d > \frac{1}{W_m} \\ \frac{W_m}{p_d}, & \text{otherwise.} \end{cases} \tag{5.130}$$

Since burst losses for fast flows trigger only TOs, substituting for all the terms in Eq. 5.12 gives the throughput for the cases of $W_m > 1/p_d$ and $W_m \le 1/p_d$.

For a medium TCP flow, the probability of triggering a TO event denoted by p_{TD} has to be obtained. It is assumed that the packets from a window of size $E[W_X]$ are distributed over an average of W_b bursts where $W_b = \frac{E[W_X]}{S}$. Since TCP SACK exits fast recovery only after all the packets in a window are acknowledged, none of the $W_b - 1$ bursts trigger a TO event and the average number of remaining bursts that experience contention but do not trigger a TO event is given by $(W_b - 1)p_c$. So p_{TD} can be obtained as [48]

$$p_{TD} = \frac{p_c}{(\frac{E[W_X]}{S} - 1)p_c + 1} \tag{5.131}$$

so that $E[W_X]$ can be written as

$$E[W_X] = \frac{8}{3} + \frac{4}{3}\sqrt{4 - \frac{3}{2}(S - \frac{S}{p_c})}. \tag{5.132}$$

Since TO can be triggered when all the bursts in a sending round are lost, Q can be written as $Q = p_d^{(\frac{E[W_X]}{S}-1)}$ (see Eq. 5.52). Using all these equations for a small value of p_c and $p_d \leq p_c$, the throughput can be approximated for a medium source by

$$B^m = \frac{1}{\text{RTT}}\sqrt{\frac{3S}{2p_c}}. \tag{5.133}$$

Another approach to improve the performance of TCP is to decrease loss of TCP ACKs in the reverse path. Usually it is assumed that the ACKs are not lost which is not realistic given the fact that ACKs are also assembled into a burst in the reverse path and a burst might contain several ACKs. A simple way to improve TCP performance by reducing the loss of ACKs is to reduce the number of ACKs in a burst or by assembling ACKs in a separate burst and give it a higher priority in scheduling at a core node. These two approaches showed good improvement of TCP throughput and the approach of assembling ACKs in a separate burst with higher priority yielded better results than assembling ACKs along with other data packets [50]. Limiting the number of ACKs in a burst prevents the simultaneous loss of many ACKs which causes a time-out. This is implemented by maintaining a separate queue at the ingress node for ACKs and data packets and ensuring that the ratio of the number of ACKs in a burst to the number of data packets does not exceed a predefined threshold. It was found that fixing the ratio to 0.4 yielded good results since it prevents loss of several ACKs as well as delaying the ACKs to a longer time [49, 50].

5.6.3 Drop Policy

When two bursts contend for a wavelength at a given time, one of the bursts is dropped. Decision of which burst needs to be dropped is usually based on a priority value fixed according to a policy. As mentioned earlier, retransmission of bursts improves the throughput of TCP. Burst-level retransmission not only increases the load in the network but also increases the percentage of bandwidth being used by dropped bursts. Hence, to avoid wastage of bandwidth in a retransmission-based loss recovery scheme, a drop policy that favors retransmitted bursts over fresh bursts (bursts transmitted for the first time) is desired. Retransmission count-based drop policy (RCDP) drops the bursts that have been transmitted fewer number of times in case of contention. This recovers the TCP packets lost in a window faster than a drop policy which is insensitive to retransmission count and based on other QoS parameters. This simple drop policy is shown to improve the throughput of TCP compared to the drop policy that always drops the burst that contends with the burst in transmission [21].

In RCDP, when an ingress node assembles a burst retransmission count (RC) is set to 1 in the corresponding BHP. When two bursts contend at a core node, the burst with lower RC field is dropped. If the RC value of two bursts is the same, the burst that arrives later is dropped. For the dropped burst a negative ACK (NAK) is sent to the ingress node so that the dropped burst is retransmitted again after increasing the RC value by one.

Note that in RCDP policy retransmission of bursts improves the performance of TCP at the same time increases the load in the network and thus the BLP. However, it reduces the time taken for the recovery of lost TCP packets since the end-to-end delay in OBS network is lower than that at the TCP level. To evaluate the performance of RCDP, it is assumed that packet loss occurs only in OBS network and not in the access networks. The time taken for successful burst transmission in OBS network in presence of burst-level retransmission (assuming that it is retransmitted n times) can be written as

$$T = t_o + \frac{L}{B} + \sum_{i=1}^{n-1}\sum_{j=I}^{c_i-1} t_p + \sum_{j=c_i}^{I-1} t_{dp} + t_w(i), \qquad (5.134)$$

where t_o is the offset time, L/B is the transmission time, $t_w(i)$ is the time for which a burst waits at the ingress node before it is retransmitted, c_i is the node where the burst is dropped, and t_p and t_{dp} are the processing times of a BHP and NAK message, respectively.

Assuming p to be the BLP, the mean BLP in RCDP is computed as follows. Suppose two bursts with RC= X and RC= Y contend and without loss of generality assume that $X < Y$. Let p_k denote the BLP when $X = k$ which is written as

$$p_k = pP[Y > k]P[X = k] + P[Y = k]P[X = k], \qquad (5.135)$$

$$= p.\frac{2N - 2k + 1}{2N^2}, \text{ for } 1 \le k \le N. \qquad (5.136)$$

The mean BLP is given by $q = \sum_{k=1}^{N} p_k = p/2$ which is same as the mean BLP with simple drop policy where one of the two contending bursts is dropped. The difference is that RCDP considers the retransmission count which causes a difference when the performance of TCP is considered as well as the load in the network. To see the difference between RCDP and simple drop policy with retransmissions, denoted by DP_r, the mean delay involved in each unsuccessful retransmission can be compared by

$$\overline{T_r} = \begin{cases} \sum_{i=1}^{H}(t_{dp} + t_p)i\frac{p}{2}(1 - \frac{p}{2})^{i-1}, & \text{for } DP_r \\ \frac{1}{N}\sum_{k=1}^{N}\sum_{i=1}^{H}(t_{dp} + t_p)ip_k(1 - p_k)^{i-1}, & \text{for RCDP.} \end{cases} \qquad (5.137)$$

It can be seen from Eq. 5.135 that as the retransmission count increases the mean BLP increases and thus the mean delay in each unsuccessful transmission. This increases the waiting time for a burst at the ingress node and the end-to-end delay (see Eq. 5.134).

Given the number of hops H along a path, the mean number of transmissions required to send a burst to the egress node is given by

$$E[N] = \sum_{k=1}^{N} k(1-r)^H (1-(1-r)^H)^{k-1}, \qquad (5.138)$$

where r is the BLP which is $p/2$ for DP_r and p_k for RCDP.

Combining Eqs. 5.137 and 5.138 gives the total mean transmission time for each burst to be delivered successfully as

$$E[T_{ie}] = Ht_p + (\overline{T_r} + \frac{L}{B})(E[N] - 1) + \frac{L}{B}. \qquad (5.139)$$

Using the mean delay for a burst given above the average RTT and RTO can be obtained for a TCP connection which can be substituted in the expression for the throughput in Eq. 5.12. RCDP prevents repeated drop of bursts (which are already being retransmitted) which ultimately leads to better throughput. Compared to dropping bursts without considering retransmission count, this approach leads to a lower BLP and therefore better throughput.

References

1. Bharat, K., Vokkarane, V.M.: Source-ordering for improved TCP performance over load-balanced optical burst-switched (OBS) networks. In: Proceedings of BROADNETS, pp. 234–242 (2007)
2. Bimal, V., Venkatesh, T., Murthy, C.S.R.: A Markov chain model for TCP NewReno over optical burst switching networks. In: Proceedings of IEEE GLOBECOM, pp. 2215–2219 (2007)
3. Bimal, V., Venkatesh, T., Murthy, C.S.R.: A stochastic model for TCP over optical burst switching network. Tech. Report, CSE Dept., IIT Madras, India, 2009
4. Bose, S.K.: An Introduction to Queueing Systems. Kluwer Academic/Plenum Publishers, USA (2002)
5. Brakmo, L.S., Peterson, L.L.: TCP Vegas: End to end congestion avoidance on global Internet. IEEE Journal on Selected Areas in Communications 13(8), 1465–1480 (1995)
6. Cameron, C., Vu, H.L., Choi, J., Bilgrami, S., Zukerman, M., Kang, M.: TCP over OBS – fixed-point load and loss. Optics Express 13(23), 9167–9174 (2005)
7. Cao, X., Li, J., Chen, Y., Qiao, C.: Assembling TCP/IP packets in optical burst switched networks. In: Proceedings of IEEE GLOBECOM, pp. 84–90 (2002)
8. Casoni, M., Raffaelli, C.: Analytical framework for end-to-end design of optical burst-switched networks. Optical Switching and Networking 4(1), 33–43 (2007)
9. Casoni, M., Raffaelli, C.: TCP performance over optical burst-switched networks with different access technologies. Journal of Optical Communications and Networking 1(1), 103–112 (2009)

10. Christodoulopoulos, K., Varvarigos, E., Vlachos, K.G.: A new burst assembly scheme based on the average packet delay and its performance for TCP traffic. Optical Switching and Networking **4**(2), 200–212 (2007)
11. Detti, A., Listanti, M.: Impact of segments aggregation of TCP Reno flows in optical burst switching networks. In: Proceedings of IEEE INFOCOM, pp. 1803–1812 (2002)
12. Detti, A., Listanti, M.: Amplification effects of the send rate of TCP connection through an optical burst switching network. Optical Switching and Networking **2**(1), 49–69 (2005)
13. de Dios, O.G., de Miguel, I., Lopez, V.: Performance evaluation of TCP over OBS considering background traffic. In: Proceedings of Optical Network Design and Modeling (2006)
14. Du, P., Abe, S.: TCP performance analysis of optical burst switching networks with a burst acknowledgment mechanism. In: Proceedings of 10th Asia-Pacific Conference on Communications and 5th International Symposium on Multi-Dimensional Mobile Communications, pp. 621–625 (2004)
15. Dunaytsev, R., Koucheryavy, Y., Harju, J.: TCP NewReno throughput in the presence of correlated losses: The slow-but-steady variant. In: Proceedings of IEEE Global Internet Symposium in conjunction with INFOCOM (2006)
16. Floyd, S.: High Speed TCP for large congestion windows. RFC 3649, December 2003
17. Gowda, S., Shenai, R., Sivalingam, K., Cankaya, H.C.: Performance evaluation of TCP over optical burst switched WDM networks. In: Proceedings of IEEE ICC, pp. 39–45 (2003)
18. Guidotti, A.M., Raffaelli, C., de Dios, O.G.: Effect of burst assembly on synchronization of TCP flows. In: Proceedings of BROADNETS, pp. 29–36 (2007)
19. Hong, D., Pope, F., Reynier, J., Baccelli, F., Petit, G.: The impact of burstification on TCP throughput in optical burst switching networks. In: Proceedings of International Teletraffic Congress, pp. 89–96 (2003)
20. Jayaraj, A., Venkatesh, T., Murthy, C.S.R.: Loss classification in optical burst switching networks using machine learning techniques: Improving the performance of TCP. IEEE Journal on Selected Areas in Communications **26**(6), 45–54 (2008)
21. Lee, S.K., Kim, L.Y.: Drop policy to enhance TCP performance in OBS networks. IEEE Communications Letters **10**(4), 299–301 (2006)
22. Malik, S., Killat, U.: Impact of burst aggregation time on performance of optical burst switching networks. Optical Switching and Networking **2**, 230–238 (2005)
23. Misra, V., Gong, W., Towsley, D.: A fluid-based analysis of a network of AQM routers supporting TCP flows with an application to RED. In: Proceedings of ACM SIGCOMM, pp. 151–160 (2000)
24. Padhye, J., Firoiu, V., Towsley, D., Kurose, J.: Modeling TCP throughput: A simple model and its empirical validation. ACM SIGCOMM Computer Communication Review **28**(4), 303–314 (1998)
25. Peng, S., Li, Z., He, Y., Xu, A.: TCP window-based flow-oriented dynamic assembly algorithm for OBS networks. Journal of Lightwave Technology **27**(6), 670–678 (2009)
26. Perello, J., Gunreben, S., Spadaro, S.: A quantitative evaluation of reordering in OBS networks and its impact on TCP performance. In: Proceedings of Optical Network Design and Modeling, pp. 1–6 (2008)
27. Raffaelli, C., Zaffoni, P.: Simple analytical formulation of the TCP send rate in optical burst switched networks. In: Proceedings of IEEE ISCC, pp. 109–114 (2006)
28. Ramantas, K., Vlachos, K.G., de Dios, O.G., Raffaelli, C.: TCP traffic analysis for timer-based burstifiers in OBS networks. In: Proceedings of Optical Network Design and Modeling, pp. 176–185 (2007)
29. Schlosser, M., Patzak, E., Gelpke, P.: Impact of deflection routing on TCP performance in optical burst switching networks. In: Proceedings of International Conference on Transparent Optical Networks, pp. 220–224 (2005)
30. Shihada, B., et.al.: BAIMD: A responsive rate control for TCP over optical burst switched (OBS) networks. In: Proceedings of IEEE ICC, pp. 2550–2555 (2006)

31. Shihada, B., Ho, P.H.: A novel TCP with dynamic burst contention loss notification over OBS networks. Computer Networks **52**(2), 461–471 (2008)
32. Shihada, B., Ho, P.H.: Transport control protocol in optical burst switched networks: Issues solutions, and challenges. IEEE Communications Surveys and Tutorials **Second Quarter**, 70–86 (2008)
33. Shihada, B., Ho, P.H., Zhang, Q.: A novel false congestion detection scheme for TCP over OBS networks. In: Proceedings of IEEE GLOBECOM, pp. 2428–2433 (2007)
34. Shihada, B., Ho, P.H., Zhang, Q.: TCP-ENG: Dynamic explicit congestion notification for TCP over OBS networks. In: Proceedings of IEEE ICCCN, pp. 516–521 (2007)
35. Shihada, B., Ho, P.H., Zhang, Q.: A novel congestion detection scheme in TCP over OBS networks. Journal of Lightwave Technology **27**(4), 386–395 (2009)
36. Shihada, B., Zhang, Q., Ho, P.H.: Performance evaluation of TCP Vegas over optical burst switched networks. In: Proceedings of BROADNETS, pp. 1–8 (2006)
37. Shihada, B., Zhang, Q., Ho, P.H.: Threshold-based TCP Vegas over optical burst switched networks. In: Proceedings of IEEE ICCCN, pp. 119–124 (2006)
38. Tijms, H.C.: A First Course in Stochastic Models. John Wiley & Sons, USA (2003)
39. Venkatesh, T., Praveen, K., Sujatha, T.L., Murthy, C.S.R.: Performance evaluation of high speed TCP over optical burst switching networks. Optical Switching and Networking **4**(1), 44–57 (2007)
40. Venkatesh, T., Sankar, A., Jayaraj, A., Murthy, C.S.R.: A complete framework to support controlled retransmission in optical burst switching networks. IEEE Journal on Selected Areas in Communications **26**(3), 65–73 (2008)
41. Venkatesh, T., Sujatha, T.L., Murthy, C.S.R.: A novel burst assembly algorithm for optical burst switched networks based on learning automata. In: Proceedings of Optical Network Design and Modeling, pp. 368–377 (2007)
42. Vlachos, K.G.: Burstification effect on the TCP synchronization and congestion window mechanism. In: Proceedings of BROADNETS, pp. 24–28 (2007)
43. Yang, Y.R., Lam, S.S.: Generalized AIMD congestion control. In: Proceedings of International Conference on Network Protocols, pp. 187–198 (2000)
44. Yu, X., Chen, Y., Qiao, C.: Performance evaluation of optical burst switching with assembled burst traffic input. In: Proceedings of IEEE GLOBECOM, pp. 1803–1812 (2002)
45. Yu, X., Qiao, C.: TCP performance over OBS networks with multiple flows input. In: Proceedings of BROADNETS, pp. 1–10 (2006)
46. Yu, X., Qiao, C., Liu, Y.: TCP implementations and false timeout detection in OBS networks. In: Proceedings of IEEE INFOCOM, pp. 358–366 (2004)
47. Yu, X., Qiao, C., Liu, Y., Towsley, D.: Performance evaluation of TCP implementations in OBS networks. Tech. Report, CSE Dept., SUNY Buffalo, 2003
48. Zhang, Q., Vokkarane, V.M., Wang, Y., Jue, J.P.: Analysis of TCP over optical burst switched networks with burst retransmission. In: Proceedings of IEEE GLOBECOM, pp. 1978–1983 (2005)
49. Zhou, J., Wu, J., Lin, J.: Improvement of TCP performance over optical burst switching networks. In: Proceedings of Optical Network Design and Modeling, pp. 194–200 (2007)
50. Zhou, J., Wu, J., Lin, J.: Novel acknowledgement-based assembly and scheduling mechanisms in optical burst switching grid networks. Optical Engineering **46**(6), 1–6 (2007)
51. Zhu, L., Ansari, N., Liu, J.: Throughput of high-speed TCP in optical burst switched networks. IEE Proceedings on Communications **152**(3), 349–352 (2005)

Index

T. Venkatesh, C. Siva Ram Murthy, *An Analytical Approach to Optical Burst Switched Networks*, DOI 10.1007/978-1-4419-1510-8,
© Springer Science+Business Media, LLC 2010